Anderson Assaid Simão
Angelita Duarte Corrêa

Cassava Leaves: Antioxidant Compounds and Fatty Acid Profile

Anderson Assaid Simão
Angelita Duarte Corrêa

Cassava Leaves: Antioxidant Compounds and Fatty Acid Profile

Characterization of antioxidant compounds and fatty acids from different cultivars and at various plant ages

ScienciaScripts

Imprint

Cover image: www.ingimage.com

This book is a translation from the original published under ISBN 978-613-9-62776-9.

Publisher:
Sciencia Scripts
is a trademark of
Dodo Books Indian Ocean Ltd. and OmniScriptum S.R.L publishing group

120 High Road, East Finchley, London, N2 9ED, United Kingdom
Str. Armeneasca 28/1, office 1, Chisinau MD-2012, Republic of Moldova, Europe
Printed at: see last page
ISBN: 978-620-7-75066-5

SUMMARY

1. INTRODUCTION

Since the world is going through a process of population explosion and major changes in climatic conditions, two aspects have been a cause for concern: the need for a constant increase in food production and the awareness of the population to avoid waste, especially with parts of food that are not traditionally used much.

In markets, fairs and distributors of plant products, the leaves and stalks of most foods are discarded. These include cassava leaves (*Manihot esculenta* Crantz).

These leaves are a viable solution to the growing demand for food, as they are rich in protein, vitamins and minerals at a low cost compared to conventional leafy vegetables. In addition, their use can provide extra income for many producers who live off cassava cultivation.

The search for natural substances that can bring benefits, especially to health, and adding value to this agricultural by-product, can contribute to greater use of cassava leaves in food and in a wide variety of areas.

Experimental and epidemiological studies have shown that people who consume foods rich in antioxidants and polyunsaturated fatty acids may have reduced their risk of numerous diseases, such as cancer, cardiovascular diseases, chronic diseases and ageing, among others.

Antioxidants are substances that fight free radicals, which are extremely reactive species that cause the oxidation of various biomolecules present in our bodies, causing various pathologies, such as cancer, neurodegenerative disorders, cardiovascular diseases and other chronic diseases, and may be the cause or aggravation of their general condition. In addition, with the problems encountered with synthetic antioxidants, which are used in food preservation and which, in addition to being expensive to produce, have also been shown by toxicological studies to cause undesirable effects in the human and animal organism, the search for natural substances that are efficient sources of antioxidants, at a lower cost than synthetic antioxidants, has been the subject of several studies.

Recently, research has been carried out with great medical and nutritional interest on long-chain polyunsaturated fatty acids from the omega-3 and omega-6 families. According to several studies, various diseases, such as cancer, arthritis and diabetes, are related to the disproportion between omega-6 and omega-3 acids. For many generations, the ratio between these fatty acids has remained at 1:1, due to the consumption of vegetables (seeds, leaves and roots) and marine products, which have adequate concentrations of omega-3. However, changes in eating habits, the emergence of processed foods and the hydrogenation of vegetable oils have reduced omega-3 concentrations, increasing the omega-6 content in our diets, changing the ratio to 10:1 to 20:1, which has had drastic consequences for health. Consequently, it is of fundamental importance to carry out studies to obtain new sources of omega-3 fatty acids in order to re-establish the ratio between the sum of omega-6/omega-3 acids.

Studies on leafy greens have shown that they contain considerable amounts of vitamins, carotenoids, phenolic compounds, among others. These compounds are one of the main defense mechanisms in the fight against free radicals and are powerful antioxidants. It is therefore important to evaluate their composition, bearing in mind that cassava leaves, until now considered to be agricultural by-products, could be a good source of these substances, which would arouse great interest in the pharmaceutical, food and cosmetics industries, leading to the valorization of this agricultural by-product.

Chlorophyll, the photosynthetic pigment in plant leaves, is directly related to their nutritional status. The amount of this pigment has been used as an index to assess the nutritional status of various types of crops. Thus, variables such as chlorophyll and antioxidant levels may be directly correlated, and studies are needed to confirm this correlation which, if confirmed, could serve as a parameter for obtaining antioxidant substances in cassava leaves.

Therefore, this work was carried out with the aim of quantifying antioxidant substances, chlorophyll and fatty acid composition and measuring antioxidant activity in cassava leaf meal from different cultivars at various ages of the plant.

2. THEORETICAL FRAMEWORK

2.1 General considerations

Cassava (*Manihot esculenta* Crantz) is a shrubby perennial plant belonging to the Euphorbiaceae family and the Manihot genus (Lorenzi, 2003). Cassava, aipim and macaxeira are some of its common names. Originally from the American continent, probably central Brazil, manioc was already widely cultivated by the Aborigines at the time of the discovery of Brazil. They were responsible for spreading it throughout most of America and the Portuguese and Spanish for spreading it to other continents, especially Africa and Asia.

Cassava is a type of crop grown in arid climates, has the ability to grow in poor soils and is relatively resistant to weeds and insects; it resists drought well and is widely adapted to the most varied climate and soil conditions, characteristics that make it cultivated all over the world, especially in poor regions (Lorezi & Dias, 1993).

With a root production of 26.2 million tons in 2009 and a similar estimate for 2010, Brazil is the world's largest cassava producer. The states of Parà, Bahia, Paranà, Maranhao and Rio Grande do Sul stand out as the largest producers, accounting for 60.5% of Brazilian production (Instituto Brasileiro de Geografia e Estatistica - IBGE, 2010). In the world, cassava is also one of the main agricultural crops and is of great social importance as the main source of carbohydrates for more than 500 million people, essentially in developing countries (Empresa Brasileira de Pesquisa Agropecuaria - EMBRAPA, 2010).

Cultivated in all regions of the country, cassava plays an important role in generating jobs and income, especially for small and medium-sized producers, who use it to feed animals and humans. The most important part of the plant are the tuberous roots, rich in starch, which are mainly destined for human consumption. It is used for the preparation of numerous derivatives, such as zest, zest flour, starch, beiju, sago, tapioca, table flour, etc. It is also used in the food industry or as a raw material in other industries (Otsubo, 2004).

In recent years, emphasis has been placed on the aerial part of the plant, which until now has been treated as an agricultural by-product, but which, nutritionally, has great potential for human and animal consumption (Lorenzi, 2003). From a nutritional point of view, only the upper third of the aerial part is used, which is represented by the leaves (Carvalho & Kato, 1987).

According to Cereda (2001), 2.5 tons of cassava leaves are produced per ton of root^{-1} . As Brazil's cassava production in 2009 was 26.2 million tons, approximately 65.50 million tons of cassava leaves were wasted that year alone.

In some Brazilian regions, such as the northeast and the north of Minas Gerais, where there is a shortage of food for the population, cassava leaves have been used for human consumption to combat hunger and protein malnutrition. The leaf has been used as one of the constituents of the so-called multimixture, increasing school meals and supplied to needy families along with food baskets.

On the African continent, cassava leaves are eaten as a vegetable and form part of the regular diet (Gidamis et al., 1993).

The use of cassava leaves in human and animal feed is justified, as they are rich in protein, with levels ranging from 20.77 to 35.9 g 100 g^{-1} MS (Ortega-Flores et al., 2003; Corrêa et al., 2004; Wobeto et al., 2006), 2006); vitamins, with levels ranging from 14.09 to 257.64 mg 100 g^{-1} MS (Corrêa et al., 2004; Wobeto et al., 2006) and minerals, especially Mg, Fe, Zn and Mn (Wobeto et al., 2006) at a low cost when compared to conventional leafy vegetables.

Although cassava leaves are nutritious, they also contain some substances considered to be anti-nutritive and/or toxic that can compromise their use, such as cyanide, polyphenols (tannins), nitrate, oxalic acid, hemagglutinin, saponins and trypsin inhibitors. However, some of these substances, depending on the amount consumed or the circumstance, can have health benefits (Melo et al., 2007; Wobeto et al., 2007).

As one of the goals of this work is to add value to cassava leaves, the following are the substances that have been the subject of several studies, as they provide various health benefits: antioxidants and the essential fatty acids omega-3 and omega-6.

2.2 Free radicals and antioxidants

Free radicals are chemical species capable of independent existence, containing one or more unpaired electrons, with a very short half-life. As such, they are highly reactive and capable of attacking any biomolecule, characterizing them as destructive molecules (Halliwell, 1994; Bianchi & Antunes, 1999).

The most important radicals present in living organisms are hydroxyl radicals (HO'), superoxide anion (O_2^- *), peroxyl (ROO'), alkoxyl (RO') and nitric oxide ('NO). There are other ROS in organisms that are not free radicals, but can induce radical reactions, such as singlet oxygen (O^1_2), hydrogen peroxide (H O_{22}) and hypochlorous acid (HOCl) (Lima & Abdalla, 2001; Yildrim et al., 2001).

In fact, free radical is not the ideal term to describe these reactive agents, as some of them do not have unpaired electrons in their last layer. As most of them are derived from oxygen metabolism, the term "reactive oxygen species" (ROS) is often used to refer to them. When these reactive species have an unpaired electron on the nitrogen atom, they are called "reactive nitrogen species" (RNS) (Barreiros et al., 2006).

These radicals can be generated by endogenous or exogenous sources. Endogenous sources originate from biological processes that normally occur in the body, such as the reduction of flavins and thiols; the result of the activity of oxidases, cyclooxygenases, lipoxygenases, dehydrogenases and peroxidases; the presence of transition metals inside the cell and electron transport systems. This generation of free radicals involves various cell organelles, such as mitochondria, lysosomes, peroxisomes, nucleus, endoplasmic reticulum and membranes. Exogenous sources of free radicals include smoking, air pollution, organic solvents, anesthetics, pesticides and radiation (Yildrim et al., 2001; Soares, 2002).

Generally, metabolism balances their presence in the body, but the problem is that when they are formed in excess, under certain abnormal conditions, a situation is created which is called oxidative stress (Tavares et al., 2000; Barreiros et al., 2006). It is now known that oxidative stress is associated with various cellular toxicity processes, such as damage to the structure of proteins, lipid peroxidation in membranes, alteration of DNA and inactivation of enzymes (Bianchi & Antunes, 1999).

In this way, they are involved in various pathologies, such as diseases of the gastrointestinal tract, such as gastric and intestinal ulcers; diseases related to the nervous system, such as Down's syndrome and Alzhmeier's disease; cardiovascular diseases, such as atherosclerosis and cancer. All the damage caused to the body can lead to premature ageing and may be the cause or aggravation of these diseases (Halliwell et al., 1992).

With the discovery of the deleterious effects caused by free radicals in the human body and in food, there has been a great deal of interest in the study of substances known as antioxidants. In food, they affect nutritional quality, as well as color, aroma, taste and texture, and are also responsible for the formation of potentially toxic and anti-nutritional compounds in the body.

Antioxidants are substances capable of slowing down or inhibiting the oxidation of oxidizable substrates and their main function is to protect cellular constituents and maintain the cellular redox state (Halliwell, 2001). They prevent the deleterious effects of oxidation by inhibiting the onset of lipoperoxidation, sequestering free radicals and/or chelating metal ions; they protect aerobic organisms from oxidative stress, defined as an increase in the formation of reactive oxygen species (Rodrigues et al., 2003). Although antioxidants donate electrons to certain substrates, they do not become free radicals, as they are stable in both forms.

Research carried out in recent years has shown that many vegetables contain compounds with antioxidant action, including vitamins C, phenolic compounds, carotenoids and some minerals (Machlin & Bendich, 1987).

According to Halliwell (1996), vitamin C and phenolic compounds act as hydrophilic antioxidants, while carotenoids act as lipophilic antioxidants.

Recently, interest in natural antioxidants for use in medicine and food has increased considerably, as many studies have proven the harmful effects of free radicals on the human body and have restricted the use of synthetic antioxidants, such as butylated hydroxytoluene (BHT) and butylated hydroxy anisole (BHA), due to their mutagenic and carcinogenic effects, as well as evidence of various other harmful effects (Yildrim et al., 2001; Zheng & Wang, 2001).

Fruits, vegetables and plants in general (including leaves) are considered good sources of antioxidants, which may be more efficient and less expensive than synthetic supplements in protecting the body from oxidative damage under different conditions. Vegetable oxidants vary widely in their contents and profiles between different types of vegetables (Leong & Shui, 2002).

In order to protect the cell and body organs against free radicals, biological systems have developed a highly complex and sophisticated antioxidant system, involving a variety of enzymatic components, small molecules and the metal chelation system, which are described below.

2.2.1 Classification and antioxidant defense strategies

Due to the continuous production of free radicals during metabolic processes in the human body, many antioxidant defense mechanisms have emerged to limit intracellular levels and prevent the induction of damage (Sies, 1993).

The antioxidants that represent the body's defense against reactive species can be classified as enzymatic (endogenous) and non-enzymatic (exogenous), depending on the structure of the oxidizing agent. The enzymatic system is made up of various enzymes, including superoxide dismutase (SOD), catalase (CAT) and glutathione peroxidase (GPx); the first to act, it prevents hydrogen peroxide and radicals such as superoxide from accumulating.

SOD plays a fundamental role in the body's defense against reactive oxygen species, as it acts to remove the superoxide radical. It catalyzes the destruction of the superoxide anion radical, converting it into oxygen and hydrogen peroxide. SOD is found in the body in two forms. The first contains copper and zinc and occurs in the cytosol, and the second contains manganese and is localized in the mitochondria (Barreiros et al., 2006).

Another enzymatic antioxidant is CAT, a cytoplasmic hemeprotein that breaks down excess hydrogen peroxide, a toxic product of the body, into water and molecular oxygen (Bianchi &

Antunes, 1999). It is located in the peroxisomes and therefore has little action in organs such as the heart and brain, as they have relatively few peroxisomes. Hydrogen peroxide, when in excess, can cause, in addition to oxidative stress, oxidation of hemoglobin, which can lead to a decrease in oxygen concentrations and cause infections and ulcers, among other diseases (Barreiros et al., 2006).

As for the GPx enzyme, it is characterized as an enzymatic antioxidant because it breaks down hydrogen peroxide into water and oxygen, but it only acts in this reaction when hydrogen peroxide is present in low concentrations. It is an enzyme located in the cytosol and mitochondrial matrix and its main sites of action are the liver and erythrocytes, but it can also occur in the heart, lung and muscle. GPx has a cisterna residue containing selenium covalently linked to the rest of the enzyme, and the presence of this mineral is essential for the proper development of this enzyme (Rover Júnior et al., 2001).

Even though the body has a high defense capacity, endogenous antioxidant defenses are not infallible and do not fully protect cellular components. It is therefore well established that dietary antioxidants are indispensable for proper defense against free radicals and thus play an important role in maintaining health. In food, the main substances that stand out against the effects caused by free radicals are vitamins C, phenolic compounds and carotenoids.

2.2.2 Antioxidant substances

Vitamin C

Vitamin C, or simply ascorbic acid, is a water-soluble and thermolabile vitamin. Because it is highly soluble in water, it is located in the aqueous compartments of organic tissues (Barreiros et al., 2006). Plants and various mammals are able to synthesize it from glucose and galactose. Humans, guinea pigs and other primates are the only mammals unable to synthesize ascorbic acid, so it is considered essential in the diet. In them, a genetic deficiency of gulonolactone oxidase prevents the synthesis of L-ascorbic acid from glucose (Lehninger, 2006).

Vitamin C is found in nature in two forms: reduced or oxidized. Both are equally active, but the oxidized form is much less widespread in natural substances. It has the following functions: it prevents scurvy; it is important in the body's defense against infections and fundamental in the integrity of blood vessel walls; it is essential for the formation of collagen fibers found in practically all tissues in the human body (dermis, cartilage and bones); it plays an important role in the recycling of vitamin E and, due to its ability to interconvert between the reduced and oxidized forms, it participates in oxidation-reduction reactions.

Its action as an antioxidant (Figure 1) is due to the fact that ascorbate (1) can donate hydrogen atoms to reduce another compound (for example, ROS or NROS). Its oxidation initially produces the semi-hydroascorbate radical (2), which is not very reactive. This radical can be

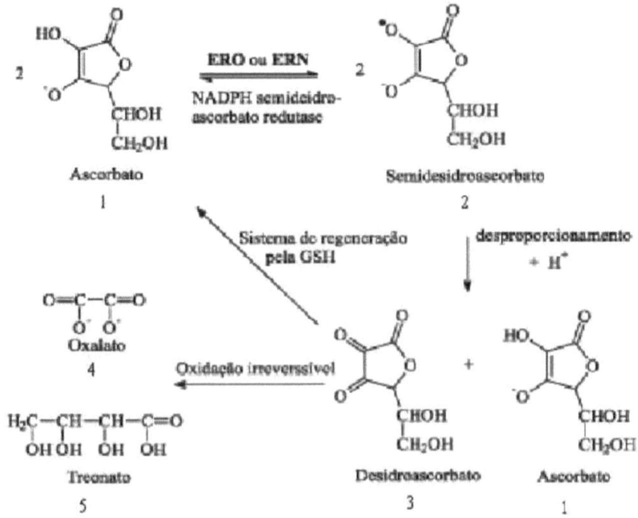

FIGURA 1 Ascorbate oxidative cycle.

Source: Barreiros et al. (2006).

The dehydroascorbate can be reconverted into ascorbate, or two molecules of it can undergo disproportionation, giving rise to a molecule of dehydroascorbate (3) and regenerating a molecule of ascorbate. Dehydroascorbate can then be regenerated into ascorbate, by means of an enzymatic system, or be irreversibly oxidized, generating oxalate (4) and threonate (5) (Halliwell & Guitteridge, 1999; Barreiros et al., 2006).

Since ascorbate converts ROS and NROS into harmless species and since ascorbate derivatives are not very reactive, it acts as an antioxidant in vivo. Vitamin C acts in the aqueous phase as an excellent antioxidant against free radicals, but is unable to act in lipophilic compartments to inhibit lipid peroxidation. Due to its antioxidant properties, many authors suggest higher daily intakes of ascorbate to protect against the development of chronic, cardiovascular diseases and some types of cancer.

Studies on cassava leaves have reported vitamin C contents ranging from 42.83 to 257.64 mg $100g^{-1}$ MS (Carvalho et al., 1989; Corrêa et al., 2004; Wobeto et al., 2006; Melo et al., 2007).

Wobeto et al. (2006) found that the vitamin C content of cassava leaves increased as the plant matured, in a study of plants aged 12, 15 and 17 months. Carvalho et al. (1989) investigated the influence of cultivar ('Guaxupé', 'Engana Ladrâo' and 'Iracema') and plant age (11 to 18 months) on the vitamin C content of cassava leaves. The highest levels of vitamin C were found at 11 months (60.10 to 60.43 mg 100 g^{-1}) and the lowest at 18 months (42.83 to 49.66 g 100 g^{-1}), with a decline observed from the 14th month after planting.

Phenolic compounds

Phenolic compounds are one of the largest classes of secondary plant metabolites, the synthesis of which does not occur in humans. Chemically, they can be defined as substances that have an aromatic ring containing one or more hydroxyl groups. The phenolic compounds present in food include phenolic acids, flavonoids, tannins and coumarins (Ribeiro & Seravalli, 2004; Moraes & Colla, 2006).

Flavonoids are the most important group of phenolic compounds and are divided into the following subgroups: anthocyanins, flavonones, flavonols, flavans and isoflavonoids. The great structural diversity of these compounds is explained by the modifications they undergo, such as hydroxylation, methylation and acylation, among others. In plants, these compounds are essential for pigmentation, growth, reproduction and resistance to pathogens, as well as being powerful antioxidants. Phenolic compounds are formed under stressful conditions, such as infections, wounds and UV radiation, among others (Moraes & Colla, 2006).

Phenolic compounds have been reported to have antioxidant activity against free radicals, which is associated with the redox properties of hydroxyl groups and their relationship with different parts of the chemical structure (Benavente-Garcia et al., 1999). Phenolic compounds have received a lot of attention in recent years, especially because they inhibit lipid peroxidation and lipoxygenase in vitro (Soares, 2002).

In general, phenolic compounds, and flavonoids in particular, have ideal structures for scavenging free radicals and are considered to be more effective antioxidants than vitamins E and C. The antioxidant activity of flavonoids depends on their structure and can be determined by five factors: reactivity as an H and electron donor agent, stability of the flavanoyl radical formed, reactivity against other antioxidants, ability to chelate transition metals and solubility and interaction with membranes (Barreiros et al., 2006). These compounds can act in both the initiation and propagation stages of the oxidative process and the intermediate products formed by their action are relatively stable due to their aromatic ring resonance (Soares, 2002; Ramalho & Jorge, 2006).

Due to their antioxidant properties, phenolic compounds make a significant contribution to the prevention of various diseases, such as cardiovascular, cancer and neurological diseases (Sànchez-Moreno, 2002; Soares et al., 2008).

Studies on cassava leaves have reported phenolic compound levels ranging from 43.37 mg g^{-1} to 106.43 mg g^{-1} (Corrêa et al., 2004; Wobeto et al., 2007). Wobeto et al. (2007), studying plants at 12, 15 and 17 months of age, found that the levels of polyphenols in cassava leaves increased as the plant matured. The drying temperature of the leaves for the production of cassava leaf flour is another factor that affects polyphenol levels, and it has been observed that higher temperatures lead to lower levels (Padmaja, 1989; Corrêa, 2000).

Carotenoids

Carotenoids are colored substances widely distributed in nature, especially in plants, which are found in chloroplasts, always accompanying chlorophylls (Ribeiro & Seravalli, 2004). In most plants, carotenoids are masked by chlorophyll, which is found in greater quantities. However, as plants mature or age, there is a change in color caused by the disappearance of chlorophyll and, during this period, the carotenoids appear more clearly.

Carotenoids are divided into carotenes and xanthophylls. Carotenes are made up of carbon and hydrogen and xanthophylls of hydrocarbons with oxygenated functional groups. β-carotene and lycopene are examples of carotenes, while lutein and zeaxanthin are xanthophylls (Moraes & Colla, 2006).

Chemically, carotenoids are fat-soluble, polyunsaturated, tetraterpenic substances made up of eight isoprene units. Figure 2 shows the chemical structure of the most common carotenoids.

These compounds are natural colorants in fruits, vegetables, roots, birds, certain fish, crustaceans and some microorganisms. These pigments, which

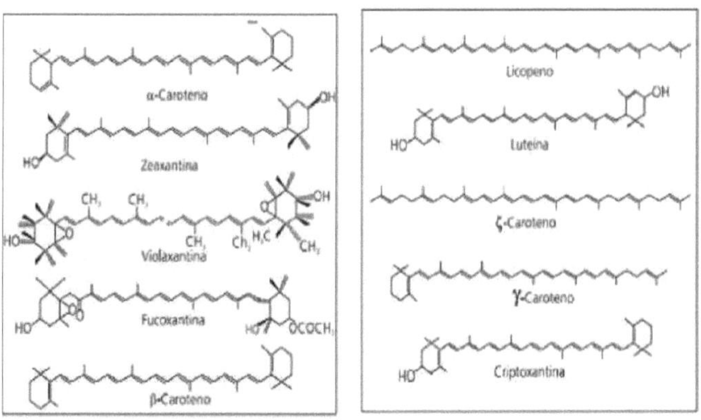

FIGURA 2 Chemical structures of carotenoids.

Source: Ambrósio et al. (2006).

from yellow to red, have aroused the curiosity of scientists since the emergence of organic chemistry, due to their relevant functions and actions (Silva & Mercadante, 2002). Only plants, bacteria, fungi and algae are capable of synthesizing carotenoids, but many animals incorporate them through their diet.

The beneficial effects of carotenoids against cancer, heart disease and macular degeneration have been recognized and have stimulated intense research into the role of these compounds as antioxidants (Uenojo et al., 2007).

Carotenoids act in vivo to combat free radicals, acting as deactivators of singlet and triplet oxygen or as peroxyl radical scavengers. They are able to stop free radical reactions that can oxidize unsaturated lipids and protect DNA against free radical attacks. As a result, they have the ability to reduce the risk of cardiovascular disease and cancer. The ability of carotenoids to exterminate free radicals is due to their structure containing double bonds, which have the ability to accommodate charges or unpaired electrons.

Carotenoids act as deactivators of triplet and singlet oxygen because their structure allows them to absorb the energy presented by these compounds during their formation, converting them into their basic forms, thus preventing the formation of free radicals and the various diseases caused by them.

Peroxyl radicals, on the other hand, are added to the double bonds of the carotene chain, generating the highly resonance-stabilized radical, which is thus unable to cause damage to other molecules (Figure 3) (Barreiros et al., 2006).

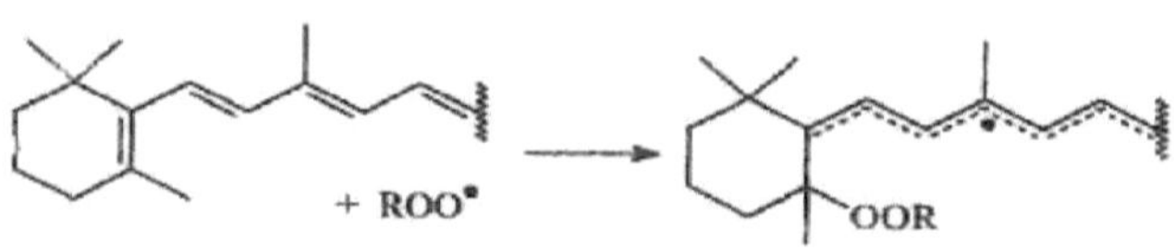

FIGURA 3 Diagram of peroxyl radical inhibition by carotenoids. Source: Barreiros et al. (2006).

Studies on cassava leaves have reported β-carotene contents ranging from 14.09 to 137.38 mg $100g^{-1}$ MS (Corrêa et al., 2004; Melo et al., 2007; Wobeto et al., 2006). Leaf maturity is also a factor that influences β-carotene levels. Wobeto et al. (2006) found that the levels of β-carotene in cassava leaves decreased as the plant matured, having studied plants at 12, 15 and 17 months of age, with the highest levels being found at 12 months of age.

2.2.3 Methods for determining antioxidants

Various techniques have been used to determine antioxidant activity in vitro, in order to allow the rapid selection of substances and/or mixtures that are potentially interesting in preventing diseases (Duarte-Almeida et al., 2006). However, there is no standard method for determining the antioxidant activity of these compounds naturally present in foods because, due to the complexity of the composition of each type of food, given that antioxidants do not act separately, the possible interaction between them can make the determination of antioxidant activity individually less effective than the total antioxidant status. Therefore, there are numerous methodologies for determining antioxidant capacity and they can be subject to interference, making it necessary to use two or more techniques, as no single method can accurately reflect the total antioxidant capacity of a sample (Huang et al., 2005).

Some of these techniques determine the ability of antioxidants to sequester free radicals generated in the environment; others evaluate the efficiency of antioxidants in inhibiting lipid peroxidation by quantifying the reaction products or measuring the inhibition of oxidation.

Among the most widely used methods are those that measure the ability of an antioxidant to reduce a free radical by hydrogen or electron donation, such as the ABTS test (2,2-azinobis-3-ethylbenzothiazoline-6-sulfonate, diammonium salt). This method is based on the antioxidant's ability to sequester the ABTS radical and prevent the radical from reacting with the substrate (Campos et al., 2008). To test antioxidant activity, the method uses the antioxidant capacity test equivalent to trolox (2-carboxylic acid-6-hydroxy-2,5,7,8-tetramethylchroman), a water-soluble vitamin E analog. It is a method widely used to evaluate the free radical scavenging capacity of many natural products and is characterized as a discoloration test applicable to hydrophilic and lipophilic antioxidants, including carotenoids and flavonoids, among other antioxidant substances (Re et al., 1999).

When it comes to measuring the ability to inhibit lipid oxidation, the β-carotene/linoleic acid coupled oxidation method is widely used. The test is based on the discoloration (oxidation) of β-carotene, induced by the oxidative degradation products of linoleic acid. The loss of the orange color characteristic of the system is monitored spectrophotometrically in order to quantify the degree of inhibition of oxidation by the antioxidant (Huang et al., 2005). The presence of an antioxidant substance in the mixture protects linoleic acid from autoxidation, preventing such discoloration (Santos et al., 2003).

Due to the variety of mechanisms and sites of action, it is difficult to define the best method to use. This has made it difficult to compare results between studies, which can only be legitimate if the solvent used to extract the compounds, the concentration of the extract and the conditions of the analytical method chosen are taken into account.

2.3 Essential fatty acids

By definition, fatty acids are aliphatic carboxylic acids obtained from the hydrolysis of natural fats and oils. They are generally made up of hydrocarbon chains ranging from 4 to 36 carbon atoms. In some fatty acids, the chain is unbranched and completely saturated; in others, the chain contains one or more double bonds, and they are called monounsaturated and polyunsaturated, respectively (Lehninger, 2006).

Differences in the size, degree and position of unsaturation in their molecules give fatty acids and the compounds that contain them different physical, chemical and nutritional properties (Belitz & Grosch, 1997).

The fatty acid composition of foods is of great importance, especially the polyunsaturated fatty acids of the omega-3 (ω-3) family, in which alpha-linolenic (C18:3), eicosapentanoic (C20:5) and docosahexanoic (C22:6) fatty acids stand out, as well as omega-6 (ω-6), linoleic (C18:2) and arachidonic acid (C20:4). These are necessary to maintain cell membranes, brain function and the transmission of nerve impulses under normal conditions. They also participate in the transfer of atmospheric oxygen to the blood plasma, in the synthesis of hemoglobin and in cell division.

Fatty acids from the ω-3 and ω-6 families are called essential fatty acids because they are not synthesized by the animal or human body, but only by plants, and must therefore be consumed in the diet.

The ω-3 and ω-6 families are derived from alpha-linolenic and linoleic acids, respectively. From these essential fatty acids, other fatty acids with long chains (chain longer than 18 carbons) and/or added unsaturation can be synthesized by humans and animals. Elongation is carried out by the enzyme elongase and desaturation by the enzyme desaturase, as shown in Figure 4.

The fatty acids shown in Figure 4 cannot be interconverted and therefore belong to two distinct families (ω-3 and ω-6). In addition, their metabolism requires the same enzymes, resulting in competition between the two families. As a result, an excess of components from one fatty acid family can interfere with the metabolism of fatty acids from the other family, which reduces their incorporation into lipid tissue and alters their biological effect.

The ω-6/ω-3 ratio in the diet of people who lived in the period before industrialization was 1:1 to 2:1, due to the abundant consumption of vegetables and foods of marine origin, rich in ω-3 fatty acids.

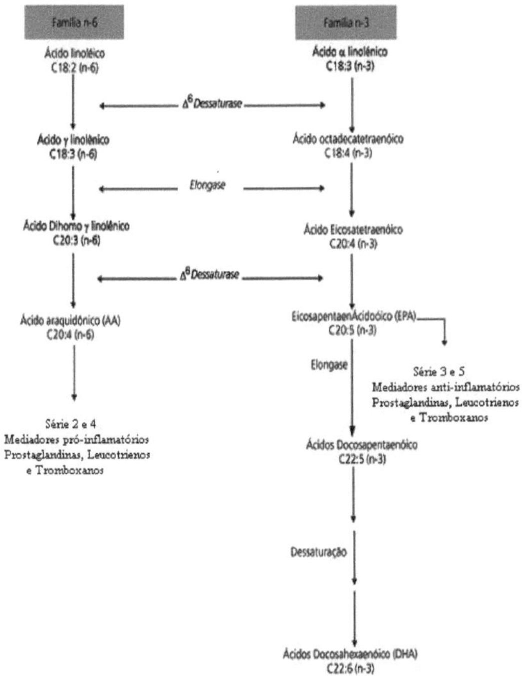

FIGURE 4 Metabolism of fatty acids from the ω-6 and ω-3 families.

Source: Iniss (2003).

With the advent of industrialization, this ratio fell sharply to between 10:1 and 20:1, mainly due to the production of refined oils from oilseed species with a high ω-6 content and a decrease in the consumption of fruit and vegetables, which resulted in diets with inadequate amounts of ω-3.

Increased intake of ω-6 is considered a problem, as it is converted to arachidonic acid, which is the precursor to the synthesis of substances known as eicosanoids, called the par series, which include prostaglandins[2] , leukotrienes[4] and thromboxanes A, whose actions, as important biochemical mediators involved in infection, inflammation, tissue damage, modulation of the immune system and platelet aggregation, are directly linked to tumor development, growth and metastasis (Wynder et al., 1997).

The ω-3 fatty acids are converted into eicosapentanoic and docosahexanoic acids, which are precursors of the chemical mediators of the odd series, called prostaglandins[3] , leukotrienes[5] and thromboxanes A (Wynder et al., 1997).

Eicosapentanoic and docosahexanoic acids block the desaturase reaction, the first step in the conversion of ω-6 fatty acids to even-series eicosanoids, as they compete with these, preventing their incorporation into the phospholipid membrane, playing an important role as an inhibitor of cell proliferation, among other damages (Johanning, 1996).

Simopoulos (2002) describes that a balanced intake of the ω-6/ω-3 ratio of 1 to 5:1 is essential, as this ratio is currently considered an index of the nutritional value of food.

Therefore, the search for alternative and complementary sources of ω-3 polyunsaturated fatty acids has become a necessity in order to establish the ratio between the sum of ω-6/ω-3 ingested.

2.3.1 Sources

Because the availability of ω-3 and ω-6 fatty acids depends on the food supply, it is important to know which foods can be a more or less valuable source of these essential fatty acids.

Linoleic and alpha-linolenic acids are found in both plant and animal species used in human food. In vegetables, alpha-linolenic acid is found in greater quantities in species with dark green leaves, as it is an important component of the polar lipid fraction contained in chloroplasts (Simopoulos, 2002).

Common sources of ω-3 fatty acids are flaxseed, vegetable oils such as linseed, canola, wheat germ and soybean. ω-6 is found in safflower, corn, soy, sunflower oil and animal fat.

ω-3 fatty acids make up more than 80% of the fatty acids found in green leafy plants. For this reason, various studies have been carried out to determine the content of ω-3 in plants. There is no information in the literature on the fatty acid composition of cassava leaves.

2.3.2 Health benefits

The intake of ω-3 fatty acids causes structural and functional changes in the phospholipid membrane. This increases the fluidity of the cell membrane, which allows for greater protein mobility and favors greater exchange of translation signals, hormone-receptor interaction and transport of substrates between the intra- and extracellular media.

There is evidence in favor of ω-3 and ω-6 fatty acids being protective against aging. They are the precursors of a group of substances with physiological and pharmacological activities called eicosanoids, which include thromboxanes, prostaglandins (which have hypotensive effects), prostacyclins (inhibit platelet aggregation and increase HDL) and leukotrienes. The balance between the production of prostaglandins and thromboxanes inhibits the onset of cardiovascular diseases (Turatti et al., 2002).

ω-3 is essential for the functioning of two important organs: the brain and the heart. In the heart, it reduces the risk of heart attacks because it prevents bad fats (hydrogenated and saturated fats) from sticking to the blood vessels and clogging them up, causing cardiovascular attacks. The brain is made up of 20% fat, so it's important to consume ω-3 to keep it active. If it is deficient, the brain functions slowly, causing a lack of memory.

Experimental studies have shown that an increase in the consumption of ω-3 fatty acids in a diet reduces blood triacylglycerol levels, heart disease, diabetes and inhibits the formation of cancers (Maillard et al., 2002; Simopoulos, 2002). Studies have also shown that the consumption of these fatty acids reduces biochemical factors associated with arthritis, acts directly on human development and has antithrombotic and anti-inflammatory actions exerted through the metabolism of eicosanoids (Simopoulos, 2002; Moraes & Colla, 2006).

2.4 Chlorophyll

Chlorophyll is the name of a group of photosynthetic pigments present in the chloroplasts of plants (in a general sense, also including algae, cyanophyceae and various protists previously considered "algae" or "plants", such as red or brown algae).

Chlorophyll's intense green color is due to its strong absorptions of the blue and red regions of the electromagnetic spectrum and, because of these absorptions, the light it reflects and transmits appears green. This pigment is able to channel the energy of sunlight into chemical energy through the process of photosynthesis. In this process, the energy absorbed by chlorophyll transforms carbon dioxide and water into carbohydrates and oxygen.

Due to their tendency to mask the presence of colors from other substances, such as carotenoids (a red or yellow pigment found in plants and animals), the predominant color in plants is green.

Chlorophyll is directly related to photosynthetic activity in plants. Therefore, the nutritional status of plants is related to the quality and quantity of chlorophyll. Several variables, such as protein content, macronutrients and micronutrients present in the leaves, can be related to the chlorophyll content present in the leaf tissue.

In the literature, the concentration of nitrogen present in corn plants and the concentration of chlorophyll are positively correlated (Castelli et al., 1996; Argenta et al., 2001). In vegetable leaves, chlorophyll and antioxidant levels may be directly correlated, but studies are needed to confirm this correlation.

To assess chlorophyll levels in plants, the Minolta SPAD-502 meter has been used. It has been investigated as a tool for quickly diagnosing the nutritional status of various crops in relation to N content, adding advantages such as simplicity of use, as well as enabling non-destructive assessment of leaf tissue (Argenta et al., 2001).

The traditional methods used to determine the amount of chlorophyll in the leaf require the destruction of tissue samples and a lot of work in the extraction and quantification processes. The development of the portable chlorophyll meter, which allows instantaneous measurements of the value corresponding to its content in the leaf without destroying it, is an alternative for estimating the relative content of this pigment (Dwyer et al., 1991; Argenta et al., 2001).

The readings taken by the portable chlorophyll meter correspond to the relative chlorophyll content of the plant leaf. The values are calculated by the equipment based on the amount of light transmitted by the leaf at two wavelengths with different chlorophyll absorbances (Minolta, 1989). The peak absorbance regions of chlorophyll are blue and red. Those with low absorbance are in the green region and those with extremely low absorbance are in the infrared region (Hendry, 1993). Because of this, the wavelengths chosen for measuring chlorophyll content or the greenness index of the leaf are in the red range, where absorbance by chlorophyll is high and unaffected by carotenoids, and in the infrared range, where absorbance is extremely low.

The chlorophyll meter is described as having two diodes that emit light at 650 nm (red) and 940 nm (infrared). The light at 650 nm is close to the two primary wavelengths associated with chlorophyll activity (645 and 663 nm). The 940 nm wavelength serves as an internal reference to compensate for differences in leaf thickness, water content or other factors (Waskom et al., 1996). The light passing through the leaf sample reaches a receiver (silicon photodiode) which converts the transmitted light into analog electrical signals. By means of an A/D converter, these analog signals are amplified and converted into digital signals, which are used by a microprocessor to calculate the SPAD values ("soil plant analysis development"), which are shown on a display. The values obtained are proportional to the chlorophyll content present in the leaf (Minolta, 1989).

3. MATERIAL AND METHODS

The work was carried out at the Department of Chemistry (Dqi) of the Federal University of Lavras (UFLA).

3.1 Harvesting, drying the leaves and preparing cassava leaf flour

Mature cassava leaves (*Manihot esculenta* Crantz) from four cultivars, 'Mocotó', 'Ufla', 'Pâo da China' and 'Ouro do Vale', obtained from the Department of Agriculture at UFLA, were harvested in the morning, at three plant ages (TIP), in the months of December (10 months), February (12 months) and April (14 months), in three replications, and transported to the Biochemistry Laboratory at DQI/UFLA. The leaves were selected intact, free of diseases and pests, washed in running water and distilled water and then placed in an air-circulating oven to dry for 48 hours at a temperature of 30°C to 35°C. After drying, they were ground (without the petioles) in a Willy-type mill and the cassava leaf meal (FFM) was stored in hermetically sealed jars in the fridge until analysis.

3.2 Moisture determination

Moisture was measured in triplicate on the fresh leaves and FFM, according to the methodology of the Association of Official Analytical Chemists - AOAC (2005), which consists of the loss of water by dehydration at temperatures of 100° to 105°C.

3.3 Determination of antioxidant substances

3.3.1 Vitamin C

a) Sample preparation

The extraction for the chromatographic analysis of vitamin C was carried out with 0.5 g of the sample in 50 mL of oxalic acid and 0.1 g of infusoria soil, shaken for 15 minutes in a horizontal shaker and then filtered through 40 g Whatman paper (Strohecker & Henningg, 1967). The extracts were filtered through a 0.22 µm Millipore membrane before being taken to the chromatograph.

b) Chromatographic analysis

A Shimadzu LC 200 A liquid chromatograph, UV-VIS detector, chromatograms at 254 nm, quaternary pump, degasser and automatic injection were used. A C_{18} Nucleosil column (250 x 4.6 mm x 5 µm) and a C_{18} pre-column (15 x 3.2 mm x 7.5 µm) were used. A pH 6.7 cap was used as the mobile phase, containing sodium acetate 0.04 mol L^{-1} , EDTA 0.05 mol L^{-1} , tributylammonium phosphate isocratically, with a flow rate of 0.6 mL min^{-1} , and a run time of 15 minutes for each sample (Silva et al., 2009). A 12-point analytical curve was carried out in triplicate with ascorbic acid ($8x10^{-6}$ to $3x10^{-3}$ mol L^{-1}) to compare the retention times and quantification of vitamin C in the FFM extracts of the four cultivars.

All the reagents used were HPLC grade and Milli-Q water.

3.3.2 Polyphenols

The extraction of phenolic compounds from FFM was carried out with 50% methanol, refluxed three times consecutively at 80°C and the extracts were pooled, evaporated to 25 mL and subjected to the dosage of phenolic compounds,

using the Folin-Denis reagent, which is reduced by phenols to a blue-colored complex in alkaline solution, which is measured at 760 nm (AOAC, 2005). The results were expressed as mg of tannic acid g^{-1} MS.

3.3.3 β-carotene

To determine β-carotene, the FFM was homogenized with a mixture of acetone and hexane (2:3) and taken for absorbance reading in a spectrophotometer at four wavelengths: (453, 505, 645 and 663 nm). The following equation was used to calculate the β-carotene concentrations:

β-carotene (mg 100 mL^{-1}) = 0.216 A_{66} 3 - 1.22 A_6 45 - 0.304 A505 + 0.452 A453.

The results were expressed in mg 100 g^{-1} of DM (Nagata & Yamashita, 1992).

3.4 Evaluation of antioxidant activity

3.4.1 Preparation of extracts

Different solvents were tested in order to choose the one that provided the greatest antioxidant activity (AA). For the tests, extracts of FFM from the Ufla cultivar at the age of 12 months were used. All the extractions were carried out in triplicate, protected from light and then submitted to the antioxidant activity detection process using the ABTS method.

The FFM was macerated in different solvents in a ratio of 1:40 (w/v) for 30 minutes and then centrifuged at 2,500 g for 15 minutes. The supernatant was collected and the precipitate was again subjected to the extraction process described above and the supernatants pooled. Nine extracting solvents were tested: 1) ethanol 50% (1st extraction)/acetone70% (re-extraction), 2) methanol 50% (1^a extraction)/acetone70% (re-extraction), 3) acetone:hexane 3:2 (1^a extraction)/ethanol70% (re-extraction), 4) methanol 50%, 5) ethanol 50%, 6) methanol p.a., 7) ethanol a.p., 8) ethyl acetate and 9) acetone.

Once the extraction solvent that resulted in the highest AA had been chosen (sub-item 4.2, Results and Discussion), the AA was assessed using the methods described below, at the three ages of the plant.

3.4.2 ABTS method

The methodology used was that developed by Re et al. (1999), with modifications made by Rufino et al. (2007). The FFM extracts were obtained according to the methodology described in sub-item 3.4.1, using ethanol 50%/acetone 70% as the extracting solvent. Four different dilutions were made of the extracts obtained for the tests and subsequent construction of the analytical curve.

The ABTS^ radical is formed by adding 5 mL of ABTS stock solution 7 mmol L^{-1} with 88 μL of potassium persulfate solution 140 mmol L^{-1} and kept in the dark for 16 hours. The radical formed was diluted in ethanol until an absorbance of 0.70±0.05 was obtained and used only on the day of analysis. From time to time, the absorbance of the diluted radical should be measured for correction, which falls rapidly, thus avoiding errors in the determination of AA.

In a dark environment, 3.0 mL of the ABTS^ radical was placed in a test tube with a 30 μL aliquot of each dilution of the extracts, homogenized in a shaker and read (734 nm) after 6 minutes of reaction, using ethanol to calibrate the spectrophotometer.

Six-point analytical curves were made with trolox (100 to 2,000 μmol L^{-1}) and ascorbic acid (10 to 200 mg mL^{-1}), as well as a test for comparison with BHT, a synthetic antioxidant, and with rutin and quercetin, which are flavonoids with proven antioxidant activity. The calculations were based on the absorbances obtained from the different dilutions (four concentrations) of the extracts, plotting a graph for each sample, with absorbance on the Y axis and the concentration corresponding to each dilution (mg L-1) on the X axis, determining the equation of the line. The absorbance corresponding to 1,000 μmol L^{-1} of the trolox standard and 1,000 mg g^{-1} of the ascorbic acid standard is then replaced in this equation, thus finding (x) the concentration of the sample (mg L^{-} 1) equivalent to 1,000 μmol of trolox L^{-1} and 1000 mg g^{-1} of ascorbic acid. The final result was calculated by dividing 1,000 (μmol L^{-1}) by the value of x (g) and multiplying by 1 g, which was expressed as μmol L^{-1} g^{-1} .

3.4.3 β-carotene/linoleic acid method

The methodology used was that developed by Marco (1968), modified by Rufino et al. (2007). The FFM extracts were obtained as described in the ABTS method.

To prepare the β-carotene/linoleic acid system solution, 50 μL of β-carotene diluted in chloroform (20 g L^{-1}) were used, to which 40 μL of linoleic acid, 530 μL of tween 20 (emulsifier) and 1 mL of chloroform were added to solubilize. The chloroform was evaporated in a rota-evaporator and 100 mL of oxygen-saturated water (distilled water treated with oxygen for 30 minutes) was added in a flask covered with aluminium to protect it from light and stirred until the system solution was yellow-orange in color. In test tubes, 5.0 mL of this system solution was added to 0.4 mL of each dilution of the sample used for the test. Control tubes were made containing 5 mL of the system solution with 0.4 mL of BHT, rutin and quercetin, together with the samples. After homogenization, readings were taken on a spectrophotometer at 470 nm, using water to calibrate the spectrophotometer, this being the zero (initial) time reading. The tubes were placed in a water bath at 40°C and sequential readings were taken every 30 minutes for a total period of 2 hours.

The percentage of oxidation inhibition was evaluated, where the reduction in absorbance of the system without antioxidant (Equation 1) is considered to be 100% oxidation.

Reduction in absorbance = Initial Abs - Final Abs (Equation 1)

The drop in the absorbance reading of the samples was correlated with the system and the percentage of oxidation was established (Equation 2). The percentage of protection against lipid oxidation is given by subtracting the percentage of oxidation of each sample from 100 (Equation 3).

$$\% \text{ Oxidation} = \frac{[(\text{Reduction Abs})\text{sample} \times 100]}{(\text{Abs reduction})\text{ system})} \text{ (Equation 2)}$$

% Protection = 100 - (% Oxidation) (Equation 3)

The antioxidant action of each sample was checked against the AA of the substances used as controls.

3.5 Chlorophyll determination

Chlorophyll was determined using a Minolta SPAD-502 device, directly on the leaf while it was still on the plant, a few moments before collection, according to the methodology described by Argenta et al. (2001). Each SPAD reading is equivalent to the result obtained from the measurement made on six leaves. Five measurements were made on each leaf, which were taken at points half to two-thirds of the length of the leaf sampled. The results were expressed in SPAD units, which are equivalent to the average of thirty readings.

3.6 Fatty acid profile

3.6.1 Lipid extraction

lipids were extracted according to the methodology of Blig & Dyer (1959). The extraction was carried out with 0.5 g of FFM in 1.5 mL of chloroform, 2.0 mL of methanol and 1.2 mL of distilled water. The resulting solution was stirred and then kept warm for 15 minutes. After this procedure, the solution was centrifuged at 2,500 g for 15 minutes. The phases were separated and the methanolic part was discarded and the chloroform packed for solvent evaporation using nitrogen gas.

3.6.2 Esterification

After evaporation of the solvent, the sample was esterified using the methodology of Joseph & Ackman (1992).

2 mL of 0.5 mol L $NaOH^{-1}$ in methanol was added to the residue obtained after evaporation. It was brought to the boil for 5 minutes and 2.5 mL of esterifying reagent was added. It was again brought to the boil for 5 minutes and then cooled. After cooling, 2.0 mL of saturated NaCl solution and 2.5 mL of hexane were added. After stirring, the sample was centrifuged at 2,500 g for 10 minutes. The phases were separated, with the lower part discarded and the upper part stored for evaporation of the solvent using nitrogen gas. The residues obtained were stored under refrigeration until the analyses were carried out.

3.6.3 Characterization of fatty acid esters

The composition of the fatty acids was determined by gas chromatography using a GC-2010 chromatograph (Shimadzu) equipped with a flame ionization detector and a 100 m long fused silica capillary column with an internal diameter of 0.25 mm, containing polyethylene glycol as the liquid stationary phase. The standard used was a mixture of 37 methyl esters (SupelcoTM 37 Component FAME Mix), from C:4 to C22:6, with a purity of 99.9%.

The following operating parameters were used: split injection mode, with a split ratio of 1:100; injection volume of 1 µL; detector temperature of 260 °C; injector temperature of 260 °C; temperature program: 4 °C/minute until reaching 140 °C, remaining at this temperature for 5 minutes, maintaining the heating ramp of 4 °C/minute until reaching 240° C, remaining at this temperature for 30 minutes.

To carry out the gas chromatography, the samples had to be redissolved in 0.50 mL of hexane.

The peaks were identified using a comparative method with the retention times of the fatty acid ester standard and the results were obtained by integrating the peak areas and expressed as a percentage of area.

3.7 Experimental design

The experiment was conducted in a completely randomized design, in a 4 x 3 factorial scheme, with four cassava cultivars and three plant ages, with three replications. To analyze the choice of solvent, the experiment was conducted with nine solvents and three replications. Statistical analysis was carried out using the Sisvar program (Ferreira, 2003), and the means were compared using the Scott-Knott test at 5% probability.

Pearson's correlation analysis was carried out between antioxidant substances, antioxidant tests and chlorophyll, using the Statistical Analysis System - SAS (1999) program, using the four cultivars and TIP as variables in three replications, and its accuracy was assessed by the correlation coefficient.

4. Results and discussion

The analysis of variance for all the parameters analyzed (Tables 1, 2 and 3, Appendix) in the cassava leaf flours (FFM) showed significant differences, at 1% or 5% probability, by the F test, except for the leaf and flour moisture variables which, in the cultivar x age interaction for both, and in the cultivar variation source for flour moisture, were not statistically different. In view of this result, the cultivar was split within each age, and then the age was split within each cultivar.

The average moisture contents (in g 100 g^{-1}) for the leaves and FFM of the four cultivars in TIP are shown in Table 1. The variations in the moisture content in g 100 g^{-1} of the leaves and FFM ranged from 63.85 to 75.45 and 5.33 to 9.67, respectively. A high moisture content was observed for cassava leaves. As there were significant differences between the cultivars and plant ages, conversions were made to dry matter for the other parameters analyzed in order to compare them across the treatments used.

3.8 antioxidant substances

3.8.1 Vitamin C

Figures 5 and 6 show the chromatograms and respective retention times of the ascorbic acid standard and a sample of FFM from the Mocotó cultivar. It can be seen that the retention peak of the sample was practically the same as that of the standard.

Table 2 shows the results of the chromatographic analysis for the quantification of vitamin C in the FFM of four cultivars, in TIP. It can be seen that the vitamin C content increased with the maturity of the plant, for

TABLE 1 Average moisture content of cassava leaves and leaf meal from four cultivates, in TIP[1] .

	Leaves (g IOOg')[1]			FFM (g IOOg')[1]		
Cultivar	10 months	12 months	14 months	10 months	12 months	14 months
Mocoto	75.45 ± 1.07 a A	72.08 ± 1.31 a A	69.48 ±2.17 a B	9.67 ± 0.35 a A	8.75 ± 3.76 a B	5.95 ± 0.22 a C
Ufla	71.51 ±0.48 b A	70.97 ± 2.83 b A	63.85 ± 0.42 b B	8.63 ± 0.52 a A	6.66 ± 0.60 a B	5.94 ±0.16 a C
Bread from China	71.75 ±0.53 a A	72.04 ± 0.75 a A	69.07 ± 2.56 a B	8.99 ± 0.40 a A	6.84 ±0.41 a B	5.33 ± 0.09 a C
Gold of the Valley	74.18 ±0.83 a A	73.22 ± 2.67 a A	68.86 ±0.66 a B	8.53 ±0.51 a A	6.60 ±0.51 a B	5.52 ± 0.47 a C

The data are the mean of three replications ± standard deviation. Lowercase letters in the columns compare the cultivars and uppercase letters in the rows compare the ages. Equal letters do not differ according to the Scott-Knott test at 5% probability;[1] TIP = three plant ages.

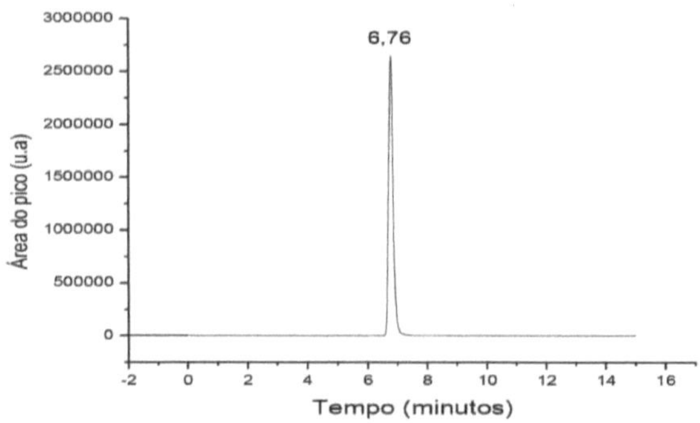

FIGURA 5 Chromatogram of the ascorbic acid standard at a concentration of 2 x 10-3 mol L^{-1} . The peak at 6.76 minutes refers to ascorbic acid.

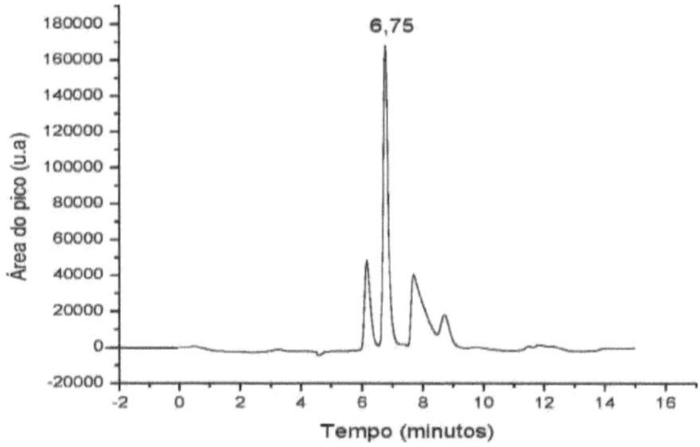

FIGURA 6 Vitamin C chromatogram of cassava leaf flour from the Mocotó cultivar at 14 months of age. The peak at 6.75 minutes corresponds to vitamin C.

TABLE 2 Average vitamin C content (mg 100 g^{-1} MS) by CLAE[1] , of cassava leaf meal from four cultivars, in TIP[2] .

Cultivar	10 months	12 months	14 months
Mocotó	281.53 ± 14.84 a B	294.77 ± 5.47 a B	521.19 ± 31.60 b A
Ufla	149.29 ± 3.84 d C	249.68 ± 1.32 c B	474.05 ± 30.07 c A
Pao da China	244.74 ± 8.58 b C	276.97 ± 11.50 b B	568.64 ± 7.71 a A
Gold of the Valley	171.44 ± 2.58 c C	310.88 ± 26.16 a B	463.69 ± 2.40 c A

Data are the mean of three replications ± standard deviation. Lowercase letters in the columns compare cultivars and uppercase letters in the rows compare ages. Equal letters do not differ according to the Scott-Knott test at 5% probability;[1] HPLC = high performance liquid chromatography;[2] TIP = three plant ages.

all the cultivars. Among the cultivars, 'Pao da China' (14 months) had the highest vitamin C content.

Carvalho et al. (1989) found an increase in vitamin C content in cassava leaves dried in an oven at 45°C up to 14 months of age. These authors also found a decline in vitamin C content between 14 and 18 months of age (starch accumulation phase). Wobeto et al. (2006), analyzing five cassava cultivars in TIP, also observed an increase in vitamin C levels as the plant matured. However, their results were contradictory to those of Carvalho et al. (1989), as vitamin C levels continued to increase between 14 and 17 months of age.

Higher levels of vitamin C were found in this study when compared to those of Wobeto et al. (2006) who, when analyzing the cultivars Mocotó and Ouro do Vale at 12 months of age, found levels of 55.72 and 64.12 mg 100 g^{-1} for Mocotó and Ouro do Vale, respectively. In the present study, however, levels of 294.77 and 310.88 mg 100 g^{-1} were found for Mocotó and Ouro do Vale, respectively. These differences may have been due to the method of analysis used, which in this study was the chromatographic method. Wobeto et al. (2006) used the colorimetric method or other factors, such as fertilization and climatic conditions, among others.

When compared to other non-conventional foods, FFM at 14 months showed average vitamin C contents higher than those found in 100 g DM of carrot leaves (203.70 mg) (Pereira et al., 2003), beet (72.42 mg) and peanuts (293.33 mg) (Sartorelli, 1998), and jaboticaba peel (246.28 mg) (Lima, 2009).

In relation to fruit, these levels are also higher than those found in jaboticaba pulp (163.38 mg 100 g^{-1} MS) (Lima, 2009) and in 100 g of fresh matter (MF) of oranges (66 mg) and papaya (149 mg) (Hernandez et al., 2006). However, they are lower than those of fruits considered to be rich in vitamin C, such as acerola, which contains 1,500 mg 100 g^{-1} of MF.

The recommended daily intake of vitamin C is 60 mg. The intake of 100 g of FFM from any cultivar at any age of the plant represents more than 100% of the daily requirement, which is a very satisfactory percentage. The cultivars therefore appear to be good sources of vitamin C with potential for use in enriching foods and cosmetics and contributing to the antioxidant potential of FFM.

4.1.2 Polyphenols

Table 3 shows the average polyphenol content of the FFM of four cultivars in TIP. It can be seen that, in general, the polyphenol content increased as the plant matured. At 14 months of age, the cultivars Pao da China and Mocotó were not significantly different and had the highest polyphenol levels. The Ouro do Vale cultivar, on the other hand, had significantly lower polyphenol levels at the three ages analyzed.

TABLE 3 Average polyphenol content (mg g^{-1} MS) of cassava leaf meal from four cultivars, in TIP[1] .

Cultivar	10 months	12 months	14 months
Mocotó	29.48 ± 0.56 a C	52.12 ± 1.13 b B	55.06 ± 0.30 a A
Ufla	29.14 ± 0.19 a C	54.53 ± 1.58a A	50.14 ± 0.66 b B
Pao da China	29.18 ± 0.30 a B	55.57 ± 1.54 a A	56.21 ± 1.39 a A
Gold of the Valley	16.46 ± 0.16 b C	31.73 ± 0.86 c B	36.26 ± 0.67 c A

Data are the mean of three replications ± standard deviation. Lowercase letters in the columns compare the cultivars and uppercase letters in the rows compare the ages. Equal letters do not differ according to the Scott-Knott test at 5% probability;[1] TIP = three plant ages.

Wobeto et al. (2007), studying the FFM of five cultivars in TIP, also observed an increase in polyphenol content as the plant matured. Analyzing different ways of drying cassava leaves (dried in the sun, in the shade and in an oven at 30°C and 4°C), Corrêa (2000) observed that, for the Baiana cultivar, the highest polyphenol contents were obtained when using the process in which the leaves were dried on laboratory benches and then placed in the sun. Consequently, the different levels of this substance can probably be attributed to the age of the plant, the cultivars and also the way the leaves were dried.

The polyphenol contents found in this study, regardless of age and cultivar, are higher than those found in mg 100 g^{-1} of MF in some vegetables, such as broccoli (68.0), onion (113.12) and cabbage (66.9) and in some fruits, such as pineapple (85.1), banana (215.7), orange (114.6), papaya (15.6) and mango (110.5) (Faller & Fialho, 2009). These figures are higher than those reported by Melo et al. (2008a), who found polyphenol contents of between 0.06 (passion fruit) and 7.9 (acerola) mg g^{-1} MF in the pulp of twelve frozen fruits. This is within the range reported by Asolini et al. (2006) who, when analyzing the phenolic content of plants used as teas, found levels ranging from 15 (yerba mate) to 56 (lemongrass) mg g^{-1} MS.

The consumption of natural antioxidants, such as the polyphenols present in most plants, has been associated with a lower incidence of diseases caused by free radicals. Consequently, the levels of this substance found in these FFM may contribute to their antioxidant capacity.

4.1.3 β-carotene

Table 4 shows the average β-carotene content of the FFM of four cultivars in TIP. It can be seen that there is an increase in β-carotene levels up to the age of 12 months and that these levels did not differ significantly from those at 14 months, except for the cultivars Ouro do Vale and Pao da China, which showed a reduction in β-carotene levels between the ages of 12 and 14 months.

TABLE 4 Average β-carotene content (mg 100 g^{-1} MS) of cassava leaf meal from four cultivars, in TIP[1] .

Cultivar	10 months	12 months	14 months
Mocotó	50.20 ± 1.60 b B	70.31 ±0.24 a A	65.92 ±2.08 a A
Ufla	53.08 ± 3.21 b B	70.07 ± 3.59 a A	65.60 ± 1.39 a A
Pao da China	65.90 ± 0.83 a B	72.72 ± 1.09 a A	61.63 ± 1.38 a B
Gold of the Valley	51.82 ± 4.59 b B	68.79 ± 4.74 a A	51.44 ± 6.67 b B

Data are the mean of three replications ± standard deviation. Lowercase letters in the columns compare the cultivars and uppercase letters in the rows compare the ages. Equal letters do not differ according to the Scott-Knott test at 5% probability;[1] TIP = three plant ages.

Wobeto et al. (2006), analyzing five cassava cultivars in TIP, also found higher levels of β-carotene at the age of 12 months. However, these authors found a decline in these levels as the plant matured. Corrêa (2000), analyzing different ways of drying cassava leaves (sun-dried, shade-dried and oven-dried at 30°C and 40°C), observed that these ways of drying resulted in significant differences in the β-carotene content, in mg 100 g^{-1} MS, in the FFM of the Baiana cultivar. The highest levels were found when the leaves were dried in an oven at 30°C (84.83) and the lowest when the leaves were dried in the shade (64.88). Consequently, the different levels of this substance can probably be attributed to the age of the plant, the cultivars and also the way the leaves were dried.

The average levels of β-carotene in the FFM studied, regardless of cultivar and plant age, ranged from 50.00 to 72.72 mg 100 g^{-1} MS and, when compared to those found in 100 g MS of other non-conventionally used green leaves, are comparable to those of sweet potato (40-120 mg) (Almazan et al., 1997), 1997), higher than that of carrots (8.70 mg) (Pereira et al., 2003) and lower than that of peanut leaves (100-140 mg) (Almazan & Begum 1996). In relation to the levels in 100 g MF of vegetables, these are higher than those of lettuce (1.37 mg), watercress (5.26 mg), spring onions (2.31 mg) and parsley (3.82 mg) (Campos et al., 2003).

As β-carotene is the precursor of vitamin A, these FFM can be considered a source of this vitamin, and these β-carotene levels can also contribute to the antioxidant potential of these flours.

4.2 Evaluation of antioxidant activity

The results of the antioxidant activity (AA) using various solvents as extractants, using the Ufla cultivar at 12 months of age, are shown in Figure 7. It can be seen that the solvent mixtures ethanol/acetone (50:70) and methanol/acetone (50:70) did not differ significantly in terms of AA and that both were superior to all the other solvents tested.

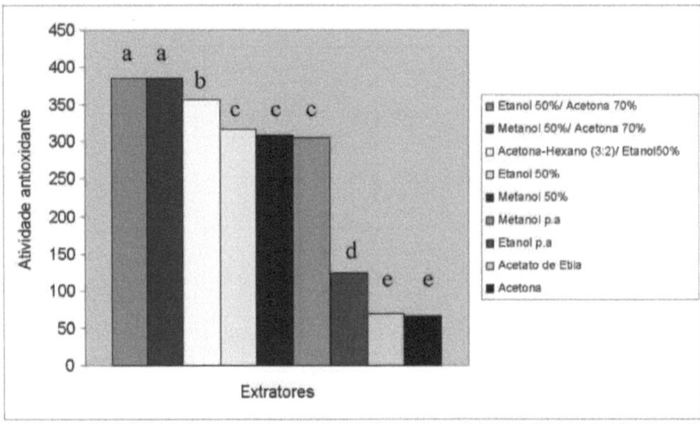

FIGURE 7 Antioxidant activity (μmol trolox L^{-1} g^{-1} of DM) of FFM, cultivar Ufla, at 12 months of age, submitted to different extractants. Equal letters indicate no difference according to the Scott-Knott test at 5% probability.

As there were no significant differences in AA between the use of ethanol/acetone (50:70) and methanol/acetone (50:70) solvents, ethanol was chosen to measure the AA of FFM in TIP, as it is less toxic and cheaper than methanol.

The selection of the ethanol/acetone solvent (50:70) proves that the use of polar and apolar solvents is more efficient in removing antioxidants, since they extract both compounds with polar characteristics (vitamin C and phenolic compounds) and apolar ones (carotenoids). It is likely that interactions between antioxidants may have effects that would not be observed when these compounds are tested in isolation.

4.2.1 ABTS method

The results of the antioxidant activity by the ABTS method, calculated by the trolox analytical curve at 6 minutes of reaction and by the standard ascorbic acid curve, of the FFM of four cultivars, in TIP, are shown in Table 5. It can be seen that AA by the ABTS method increases with plant maturity for all cultivars, both in terms of trolox equivalents and ascorbic acid. The highest AA was observed at 14 months for the Pao da China cultivar, both in trolox equivalents (680.62 μmol L^{-1} g^{-1}) and ascorbic acid (102.42 mg g^{-1}). The result shows that 1 g of FFM from this cultivar has an antioxidant capacity of 680.62 μmol trolox L^{-1} and the equivalent antioxidant power of 102.42 mg of vitamin C. The Ouro do Vale cultivar, on the other hand, had the lowest AA, both in terms of trolox equivalents and ascorbic acid, in the TIP.

The results presented by the rutin and quercetin standards demonstrate the sensitivity of the method to small changes in the structure of the molecule, since rutin has a similar structure to quercetin, but with a glycosidic substitution in one of the hydroxyls and therefore has lower AA (Duarte-Almeida et al., 2006).

When evaluating the antioxidant efficiency of FFM, regardless of plant age and cultivar, the average trolox equivalent values ranged from 155.90 to 680.62 μmol L^{-1} g^{-1} and vitamin C equivalent values ranged from 24.51 to 102.42 mg g^{-1} .

When compared to the BHT and rutin standards, the antioxidant potential of the cultivars analyzed, in terms of both trolox and ascorbic acid equivalents, is considered to be moderate, as it reached an average of almost 50% of the potential of these standards at 14 months, except for the Ouro do Vale cultivar. As for quercetin, the potential of the cultivars at any age was much lower than the antioxidant potential of this standard.

TABLE 5 Antioxidant activity of cassava leaf meal from four cultivars, in TIP[1] , and three standards, by the ABTS method.

Cultivar	μmol trolox L^{1} g^{1}			mg vitamin C g^{1}		
	10 months	12 months	14 months	10 months	12 months	14 months
Mocoto	238.62 ± 1.05 bC	320.25 ± 1.02 cB	636.12 ± 1.41 cA	38.54 ± 0.04 bC	49.11 ± 0.10 cB	100.83 ± 0.25 cA
Ufla	359.85 ± 0.86 aC	444.65 ± 0.59 aB	658.71 ± 1.84 bA	50.11 ± 0.12 aC	52.07 ± 0.20 aB	101.20 ±0.03 bA
Bread of China	240.25 ± 0.68 bC	338.15 ±0.67 bB	680.62 ± 0.30 aA	38.79 ± 0.03 bC	50.45 ±0.11 bB	102.42 ± 0.58 aA
Gold of the Valley	155.90 ± 2.39 cC	266.83 ± 0.64 dB	327.27 ± 0.08 dA	24.51 ±0.09 cC	44.60 ± 0.13 dB	50.85 ±0.18 dA
BHT		1.418,81 ±97,98			259,55 ± 3,27	
Quercetin		7.491,59 ± 119,04			942,96 ± 1,89	

Rutin	1.055,50 ± 13,39	161,29 ±0,30

The data are the mean of three replications ± standard deviation. Lowercase letters in the columns compare the cultivars and uppercase letters in the rows compare the ages. Equal letters do not differ according to the Scott-Knott test at 5% probability;[1] TIP = three plant ages.

However, as these substances are not found in isolation like the standards, the average potential of approximately 50% obtained by the flours at 14 months can be considered a good antioxidant potential. In addition, AA also depends on the synergistic effect between these compounds.

The AA in FFM exceeds that found by Kuskoski et al. (2005), in frozen fruit pulps, in µmol trolox L^{-1} g^{-1} MF: acerola (66.5), mango (11.8), grape (8.5), guava (7.2) and passion fruit (1.02); those of two grape skins in MS: 'Isabel' (89.22) and 'Niàgara' (157.31) (Soares et al., 2008), 2008) and that of agro-industrial wine residues: (98.9) (Cataneo et al., 2008). However, it was lower than that found in the peel of two jaboticaba varieties: Sabarà (2,060) and Paulista (1,560) (Lima, 2009).

4.2.2 β-carotene/inoleic acid method

Initially, the oxidation inhibition potentials of the BHT, quercetin and rutin standards were tested. The BHT standard (synthetic antioxidant) was chosen because it is widely used in the food industry, and the natural antioxidants, quercetin and rutin, because they are present in fruit and vegetables (Duarte- Almeida et al., 2006). The results (Figure 8) show that the BHT standard has a greater potential for inhibiting oxidation than quercetin, which in turn has a greater potential than rutin.

Table 6 shows the results of the inhibition of lipid oxidation by the β-carotene/linoleic acid method of FFM from four cultivars in TIP. It can be seen that the maturity of the plant had little influence on the AA of the FFM. Among the cultivars, 'Mocotó' showed the greatest potential for inhibiting oxidation in TIP.

It can be seen that when compared to the BHT standard, the Mocotó cultivar, in the TIP, showed AA similar to this standard. The other cultivars, on the other hand, had lower AA compared to BHT, but they were not that much lower

FIGURE 8 Discoloration curve of β-carotene in the presence of BHT (200 mg L-1), quercetin (200 mg L-1), rutin (200 mg L-1) and the control.

TABLE 6 Antioxidant activity of cassava leaf meal from four cultivars, in TIP1, and three standards, by the β-carotene/linoleic acid method.

	% inhibition		
Cultivar	10 months	12 months	14 months

Mocotó	93.04 ± 3.14 a A	94.05 ± 0.20 a A	89.81 ±0.19 a B
Ufla	85.17 ± 0.96 b A	86.47 ± 0.33 b A	84.61 ± 0.66 b A
Pao da China	81.47 ± 1.27 c B	85.51 ± 0.63 b A	79.79 ± 1.28 c B
Gold of the Valley	82.79 ± 0.35 c B	85.69 ± 0.80 b A	78.87 ± 2.25 c C
BHT	93,60 ± 0,56		
Quercetin	76,20 ± 1,15		
Rutin	19,21 ± 0,88		

The data are the mean of three replicates±standard deviation. Lowercase letters in the columns compare the cultivars and uppercase letters in the rows compare the ages. Equal letters do not differ according to the Scott-Knott test at 5% probability; 1TIP = three plant ages.

and the ABTS method. In relation to the quercetin and rutin standards, the cultivars in the TIP had higher AA than those of these standards. In relation to rutin, the AA of the cultivars in the TIP were much higher. Many authors report an antioxidant potential above 70% as optimal for inhibiting lipid oxidation. All the cultivars in the TIP achieved AA above 70% and therefore have a high potential for inhibiting lipid oxidation.

Factors such as the analytical method of extraction, climatic conditions, geographical origin and chemical constituents are mainly responsible for the very different results found in the literature regarding the AA of foods, and must be taken into account when comparing results in order for them to be valid.

It can be seen that the FFM AA results were moderate when the free radical scavenging method (ABTS) was used and high when the lipid oxidation inhibition method (β-carotene/linoleic acid) was used, emphasizing the importance of using more than one technique to reflect the antioxidant capacity of a sample.

4.3 Chlorophyll

The use of the SPAD-502 equipment as a chlorophyll meter is justified by the results obtained by Argenta et al. (2001) and Zotarelli et al. (2003) who, when analyzing the correlation between chlorophyll meter readings and the chlorophyll content extracted from maize leaves, obtained positive correlations in their results. Neves et al. (2005) also found a positive correlation between this relationship in herbaceous cotton leaves. Therefore, the readings taken with this equipment can replace traditional measurements of chlorophyll content in plants with good accuracy, as it is a quick, cheap and non-destructive technique.

The results of the chlorophyll readings of the FFM in TIP using the SPAD-502 portable meter are shown in Table 7.

TABLE 7 Average chlorophyll content (SPAD unit) of cassava leaf meal from four cultivars, in TIP[1] .

Cultivar	10 months	12 months	14 months
Mocotó	565.60±3.05 d A	479.80±9.07 d B	429.60±8.68 d C
Ufla	719.80±7.63 b A	631.20±5.72 b C	677.00±11.51 b B
Pao da China	589.40±10.43 c A	491.60±1.52 c B	468.40±13.67 c C
Gold of the Valley	771.00±5.20 a A	743.20±4.38 a C	758.40±0.63 a B

Data are the mean of three replications ± standard deviation. Lowercase letters in the columns compare the cultivars and uppercase letters in the rows compare the ages. Equal letters do not differ according to the Scott-Knott test at 5% probability;[1] TIP = three plant ages.

It was found that the highest chlorophyll levels were found at 10 months of age and that there was a tendency for these levels to decrease as the plant matured. The Ouro do Vale cultivar, in the TIP, had the highest chlorophyll levels, found at 10 months of age and there was a tendency for these levels to decrease as the plant matured. The Mocotó cultivar, on the other hand, had significantly lower chlorophyll levels.

The tendency for chlorophyll levels to decrease as the plant matures can probably be attributed to the leaf senescence that occurs in most plants, leading to a loss of green color and, consequently, a reduction in chlorophyll levels.

4.4 Correlation between antioxidant substances and AA and chlorophyll

The results of the correlations between the antioxidant substances (Tables 2, 3 and 4) and the AA (Tables 5 and 6) and chlorophyll (Table 7) of the FFM, in TIP, are shown in Table 8.

TABLE 8 Correlation (R) between antioxidant substances and antioxidant activity (AA) and chlorophyll of cassava leaf meal, in TIP[1] .

	R		
AA/Chlorophyll	Vitamin C	Polyphenols	β-carotene
ABTS (trolox)	0,78*	0,75*	0,28^{ns}

ABTS (vit C)	0,84*	0,70*	0,25^{ns}
β-carotene/ linoleic acid	-0,20^{ns}	0,18^{ns}	0,20^{ns}
Chlorophyll	-0,40*	-0,71*	-0,45*

* Significant at 5% probability; ns = not significant; [1] TIP = three plant ages.

It can be seen that the ABTS antioxidant tests (equivalent to trolox and vitamin C) showed a high positive correlation with vitamin C and polyphenols, but there was no correlation with β-carotene. These results indicate that vitamin C and polyphenols are the antioxidant substances that contributed most to the increase in AA in FFM, with vitamin C standing out, as it had the highest correlation coefficients, 0.78 and 0.84, for trolox and vitamin C equivalents, respectively.

The β-carotene/linoleic acid test showed no correlation with the antioxidant substances. The antioxidant substances showed a negative correlation with chlorophyll, i.e. when the levels of antioxidant substances increased, the chlorophyll levels decreased in the plant. For β-carotene, this is easily explained, because with increasing maturity, chlorophyll begins to disappear and carotenoids stand out.

Lima (2009) found a positive correlation between vitamin C and polyphenols with ABTS tests, both in terms of trolox equivalents and vitamin C, in studies carried out on jaboticaba. However, the greatest contribution to AA was made by polyphenols. Cataneo et al. (2008) and Soares et al. (2008) also observed a positive correlation between the content of total phenols and AA by the ABTS test, equivalent to trolox, in studies carried out with agro-industrial wine residues and grape skins of two varieties, respectively.

Lima (2009) found no correlation between the β-carotene/linoleic acid test and vitamin C and polyphenols in a study of jaboticaba fruit. Kubola & Siriamornpun (2008) and Melo et al. (2008b) also found no correlation between this test and polyphenol content in fruit. Kaur & Kapoor (2002) found a moderate positive correlation between polyphenol content and the β-carotene/linoleic acid test in a study of vegetables.

Duarte-Almeida et al. (2006) found that acerola, which is rich in vitamin C, showed low AA in the β-carotene/linoleic acid test, while it showed high AA in the DPPH test (free radical scavenging method), which shows that vitamin C contributes little to AA in the β-carotene/linoleic acid test.

The presence of other antioxidant substances, such as vitamin E in the extracts tested, may be contributing synergistically to the AA of FFM, and may be influencing the results of the correlations, especially in relation to vitamin C, as it acts cooperatively with vitamin E at the interface of the cell membrane in the fight against free radicals.

In addition to the presence of other phytochemicals in the extracts, the chemical structure of the active component has an influence on the efficacy of the natural antioxidant, since the position and number of hydroxyls present in the polyphenol molecules are considered to be relevant factors for AA, which can make this contribution greater for one test than for another.

4.5 Fatty acid profile

Figures 9 and 10 show, respectively, the chromatograms of the standard mixture of 37 fatty acid methyl esters and a sample of FFM from the Mocotó cultivar at 10 months of plant age. The FFM fatty acids were identified by comparing the retention times of the standards.

Tables 9, 10 and 11 show the main fatty acids (FA) present in the FFM of the four cultivars in TIP. Twenty FFA were detected in the FFM of the four cultivars in the TIP, with alpha-linolenic (C18:3n3) and palmitoleic (C16:1) being the main acids, representing an average of 33.51% and 19.30%, respectively, of the total percentage of FFA. There were practically no differences in the percentage of alpha-linolenic and palmitoleic acids at the three ages studied. It was also observed that, for other FFM FAs, changes occurred as the plant matured, with new acids appearing, while others were no longer detected, such as tridecanoic acid (C13:0) and lauric acid (C12:0).

The total amount of FA in FFM, regardless of age and cultivar, averaged 3.02% for saturates, 20.03% for monounsaturates and 33.99% for polyunsaturates. The ratio of (PUFA to SFA) averaged 11.25 and 0.04 for ω-6/ω-3. A higher percentage of unsaturated AGs than saturated AGs was observed in FFM in TIP.

Diets with an IPFA/AGS ratio of more than 0.45 (DHSS, 1984) and an ω-6/ω-3 ratio of less than 5 (Simopoulos, 2002) are considered healthy from a nutritional point of view for humans. Considering that the fat composition of the FFM in the TIP was within the recommendations, these FFM can be considered good food sources.

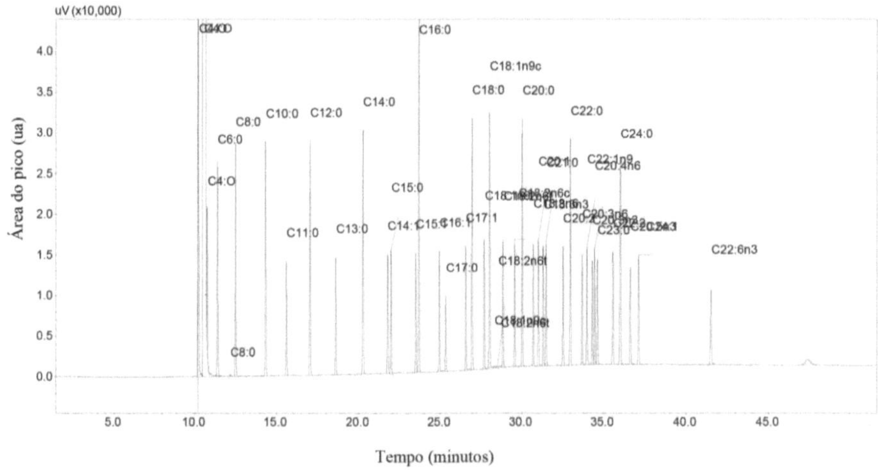

FIGURE 9 Chromatogram of the standard mixture of 37 fatty acid methyl esters.

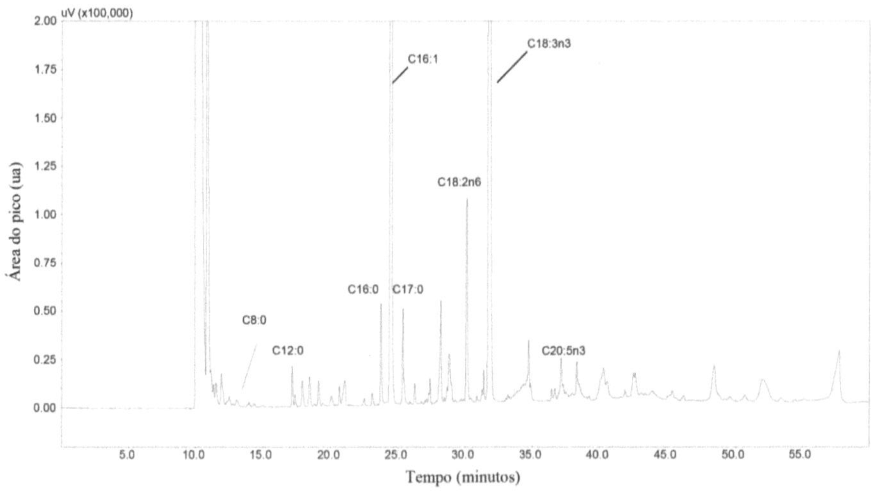

FIGURE 10 Chromatogram of fatty acids from the cassava leaf Iarinha of the Mocotó cultivar, at 10 months of age.

TABLE 9 Fatty acid composition (%) of cassava leaf meal from four cultivars at 10 months of age.

Cultivars				
Fatty acids	Mocotó	Ufla	Chinese bread	Gold of the valley
C8:0	0,45±0,21	0,24±0,16	0,42±0,35	0,15±0,01
C12:0	0,48±0,11	0,62±0,37	0,41±0,07	0,35±0,10
C13:0	nd	nd	nd	nd

C14:0	nd	nd	nd	nd
C15:1	nd	nd	nd	nd
C16:0	1,45±0,16	1,40±0,15	1,26±0,02	1,26±0,14
C16:1	22,34±z,07	18,83±0,61	20,18±0,39	19,58±0,29
C17:0	1,65±0,15	1,64±0,25	1,71±0,06	1,81±0,06
C18:0	nd	nd	0,11±0,01	nd
C18:1n9t	nd	nd	nd	nd
C18:2n6t	1,78±0,28	1,95±0,52	nd	1,69±0,28
C18:2n6c	nd	nd	nd	nd
C20:0	nd	nd	nd	nd
C18:3n6	nd	nd	nd	nd
C18:3n3	37,40±1,96	30,01±7,40	32,67±1,09	36,64±1,59
C20:3n3	nd	0,40±0,12	nd	nd
C24:0	nd	nd	0,25±0,11	nd
C20:5n3	0,31±0,06	0,18±0,04	0,49±0,29	0,15±0,05
C24:1	nd	nd	nd	nd
C22:6n3	nd	nd	nd	nd
∑ AGS	4,03	3,90	4,16	3,57
∑AGMI	22,34	18,83	20,18	19,58
∑AGPI	35,45	32,54	33,16	36,29
AGPI/AGS	8,88	8,34	7,97	10,16
ω-6 /ω-3	0,04	0,06	0,01	0,05

Data are the mean of three replicates ± standard deviation; nd = not detected; ∑ AGS = sum of saturated fatty acids; ∑AGMI = sum of monounsaturated fatty acids; ∑AGPI = sum of polyunsaturated fatty acids; AGPI/AGS = ratio between the sums of polyunsaturated and saturated acids; ω-6 /ω-3 = ratio between the sums of omega-6 and omega-3.

TABLE 10 Fatty acid composition (%) of cassava leaf meal from four cultivars at 12 months of age.

Cultivars				
Fatty acids	Mocotó	Ufla	Chinese bread	Gold of the valley
C8:0	0,21±0,09	0,16±0,06	0,28±0,07	0,19±0,08
C12:0	nd	0,31±0,14	nd	0,23±0,03
C13:0	0,49±0,08	0,69±0,04	0,53±0,08	0,59±0,15
C14:0	nd	0,29±0,12	nd	0,21±0,02
C15:1	nd	nd	nd	nd
C16:0	1,28±0,14	nd	nd	nd
C16:1	20,07±3,01	16,85±1,28	19,66±2,74	19,48±2,79
C17:0	1,92±0,04	1,76±0,25	1,68±0,25	1,81±0,14
C18:0	nd	0,16±0,00	0,12±0,01	0,18±0,09
C18:1n9t	nd	0,31±0,08	nd	0,21±0,08
C18:2n6t	2,08±0,13	2,38±0,00	nd	nd
C18:2n6c	nd	nd	nd	nd
C20:0	nd	nd	nd	0,18±0,03
C18:3n6	nd	0,35±0,01	nd	0,24±0,06
C18:3n3	35,84±0,54	30,76±6,93	28,94±5,71	33,42±0,42
C20:3n3	nd	0,15±0,07	nd	nd
C24:0	nd	nd	0,33±0,08	nd
C20:5n3	0,20±0,01	nd	0,77±0,08	0,13±0,02
C24:1	nd	0,31±0,07	nd	0,17±0,02
C22:6n3	nd	nd	nd	nd

∑ AGS	3,90	3,37	2,94	3,39
∑AGMI	20,07	17,47	19,66	19,86
∑AGPI	38,12	33,64	30,04	33,79
AGPI/AGS	9,77	9,98	10,22	9,97
ω-6 /ω-3	0,06	0,09	0,01	0,01

Data are the mean of three replicates ± standard deviation; nd = not detected; ∑ AGS = sum of saturated fatty acids; ∑AGMI = sum of monounsaturated fatty acids; ∑AGPI = sum of polyunsaturated fatty acids; AGPI/AGS = ratio between the sums of polyunsaturated and saturated acids; ω-6 /ω-3 = ratio between the sums of omega-6 and omega-3.

TABLE 11 Fatty acid composition (%) of cassava leaf meal from four cultivars at 14 months of age.

Cultivars				
Fatty acids	Mocotó	Ufla	Chinese bread	Gold of the valley
C8:0	0,31 ± 0,03	0,27 ± 0,06	0,22 ± 0,02	0,22 ± 0,01
C12:0	nd	nd	0,12 ± 0,04	nd
C13:0	0,45 ± 0,03	0,47 ± 0,08	0,54 ± 0,06	nd
C14:0	0,14 ± 0,07	0,11 ± 0,00	0,15 ± 0,00	nd
C15:1	nd	0,38 ± 0,08	0,77 ± 0,02	0,52 ± 0,09
C16:0	nd	nd	nd	nd
C16:1	21,42 ± 1,87	18,61 ± 0,47	18,15 ± 0,78	22,39 ±1,40
C17:0	nd	1,47 ± 0,69	nd	nd
C18:0	nd	0,29 ± 0,02	nd	nd
C18:1n9t	0,22 ± 0,01	0,13 ± 0,00	nd	nd
C18:2n6t	0,23 ± 0,01	0,40 ± 0,08	0,25 ± 0,04	nd
C18:2n6c	0,12 ± 0,01	0,39 ± 0,01	0,25 ± 0,05	nd
C20:0	0,25 ± 0,04	0,53 ± 0,06	0,45 ± 0,08	0,22 ± 0,15
C18:3n6	0,22 ± 0,07	0,50 ± 0,02	0,38 ± 0,02	0,16 ± 0,01
C18:3n3	32,30 ± 3,65	32,82 ± 1,06	28,67 ± 1,61	35,15 ± 0,54
C20:3n3	nd	nd	nd	nd
C24:0	0,18 ± 0,07	0,30 ± 0,04	nd	0,53 ± 0,04
C20:5n3	nd	nd	nd	nd
C24:1	nd	nd	nd	nd
C22:6n3	1,58 ± 0,49	1,47 ± 0,25	nd	nd
∑ AGS	1,08	3,44	1,48	0,97
∑AGMI	21,64	19,12	18,92	22,91
∑AGPI	34,45	35,58	29,55	35,31
AGPI/AGS	31,90	10,34	19,96	36,46
ω-6 /ω-3	0,02	0,04	0,03	0,01

Data are the mean of three replicates±standard deviation; nd = not detected; ∑ AGS = sum of saturated fatty acids; ∑AGMI = sum of monounsaturated fatty acids; ∑AGPI = sum of polyunsaturated fatty acids; AGPI/AGS = ratio between the sums of polyunsaturated and saturated acids; ω-6 /ω-3 = ratio between the sums of omega-6 and omega-3.

5. conclusions

Cassava leaves aged 14 months had the highest antioxidant levels, with the Mocotó and Pao da China cultivars standing out.

FFM are rich in antioxidant substances and have a high antioxidant capacity to protect against lipid peroxidation, but a moderate capacity to capture free radicals.

The highest antioxidant levels are found when the cassava leaf has low chlorophyll levels.

The FFMs show alpha-linolenic and palmitoleic acids as the majority, representing more than 50% of the total fatty acids. The ratios of ω-6/ω-3 and polyunsaturated/saturated acids are within the recommended levels for human consumption.

6. BIBLIOGRAPHICAL REFERENCES

ALMAZAN, A. M.; BEGUM, F.; JOHSON, C. Nutritional quality of sweetpotato greens from greenhouse plants. **Journal of Food Composition and Analysis**, Orlando, v. 10, p. 246-253, 1997.

ALMAZAN, A. M.; BEGUM, F. Nutrients and antinutrients in peanut greens. **Journal of Food Composition and Analysis**, Orlando, v. 9, p. 375-383, 1996.

AMBRÓSIO, C. L. B.; CAMPOS, F. A. C. S.; FARO, Z. P. Carotenoids as an alternative to hypovitaminosis A. **Revista de Nutriçâo**, Campinas, v. 19, n. 2, p. 233-243, 2006.

ARGENTA, G.; SILVA, P. R. F. da; FORSTHOFER, E. L.; STRIEDER, M. L. Relationship between chlorophyll meter readings and extractable chlorophyll and nitrogen content in maize leaves. **Revista Brasileira de Fisiologia Vegetal**, Londrina, v. 13, n. 2, p. 158-167, 2001.

ASOLINI, F. C.; TEDESCO, A. M.; CARPES, S. T. Antioxidant and antibacterial activity of phenolic compounds from extracts of plants used as teas. **Brazilian Journal of Food Technology**, Campinas, v. 9, n. 3, p. 209-215, jul./set. 2006.

ASSOCIATION OF OFFICIAL ANALYTICAL CHEMISTS. **Official methods of analysis of the association of the analytical chemists**. 17. ed. Washington, 2005.

BARREIROS, A. L. B. S.; DAVID, J. M.; DAVID, J. P. Oxidative stress: Relationship between the generation of reactive species and the body's defense. **Quimica Nova**, Sao Paulo, v. 29, n. 1, p. 113-123, 2006.

BELITZ, H. D.; GROSH, W. **Quimica de los alimentos**. 2. ed. Zaragoza: Acribia, 1997.

BENAVENTE-GARCIA, O.; CASTILLO, J.; LORENTE, J.; ORTUNO, A.; RIO, J. A. del. Antioxidant activity of phenolics extracted from *Olea europaea* L. leaves. **Food Chemistry**, London, v. 68, p. 457-462, 1999.

BIANCHI, M. L. P.; ANTUNES, L. M. G. Free radicals and the main antioxidants in the diet. **Revista de Nutriçâo**, Campinas, v. 12, n. 2, p. 123-130, 1999.

BLIGH, E. G.; DYER, W. J. A rapid method of total lipid extraction and purification. **Canadian Journal of Biochemistry**, Ottawa, v. 37, p. 911-917, 1959.

CAMPOS, F. M.; MARTINO, H. F.; SABARENSE, C. M.; Stability of antioxidant compounds in processed vegetables: a review. **Alimentos e Nutriçâo**, Marilia, v. 19, n. 4, p. 481-490, 2008.

CAMPOS, F. M.; SANTANA, H. M.; STRINGHETA, P. C.; CHAVES, J. B. P. Beta-carotene content in leafy vegetables prepared in commercial restaurants in Viçosa-MG. **Brazilian Journal Food Technology**, Campinas, v. 6, n. 2, p. 163-169, jul./dez. 2003.

CARVALHO, V. D.; CHAGAS, S. J. R.; MORAIS, A. R.; PAULA, M. B.

Effect of harvest time on productivity and vitamin C and β-carotene content in the aerial part of cassava cultivars (Manihot esculenta Crantz). **Revista Brasileira de Mandioca**, Cruz das Almas, v. 8, n. 1, p. 25-35, 1989.

CARVALHO, V. D.; KATO, M. S. A. Potential use of the aerial part of cassava. **Informe Agropecuàrio**, Belo Horizonte, v. 13, n. 145, p. 23-28, jan. 1987.

CASTELLI, F.; CONTILLO, R.; MICELI, F. Non-destructive determination of leaf chlorophyll content in four crop species. **Journal of Agronomy and Crop Science**, Berlin, v. 177, p. 275-283, 1996.

CATANEO, C. B.; CALIARI, V.; GONZAGA, L.; KUSKOSKI, L. M.; FETT, R. Antioxidant activity and phenolic content of agro-industrial waste from wine production. **Ciências Agrârias**, Londrina, v. 29, n. 1, p. 93-102, jan./mar. 2008.

CEREDA, M. P. **Manejo, uso e tratamento de subprodutos da mdustrialização da mandioca**. Sao Paulo: Cargill Foundation, 2001. 320 p.

(Culture of Latin American Starchy Tubers, v. 4).

CORRÊA, A. D. **Cassava leaf flour - effects of processing on some nutrients and anti-nutrients**. 2000. 108 p. Thesis (Doctorate in Food Sciences) - Federal University of Lavras, Lavras.

CORRÊA, A. D.; SANTOS, S. R.; ABREU, C. M. P.; JOKL, L.; SANTOS, C. D. Removal of polyphenols from cassava leaf flour. **Ciência e Tecnologia de Alimentos**, Campinas, v. 24, n. 2, p. 159-164, Apr./Jun. 2004.

DEPARTMENT OF HEALTH AND SOCIAL SECURITY. **Report on health and social subjects**: n. 28: diet and cardiovascular disease. London, 1984.

DUARTE-ALMEIDA, J. M.; SANTOS, R. J.; GENOVESE, M. I.; LAJOLO, F. M. Evaluation of antioxidant activity using the b-carotene/lyleic acid system and the DPPH radical scavenging method. **Ciência e Tecnologia de Alimentos**, Campinas, v. 26, n. 2, p. 446-452, 2006.

DWYER, L. M.; TOLLENAAR, M.; HOUWING, L. A non-destructive method to monitor leaf greenness in corn. **Canadian Journal of Plant Science**, Ottawa, v. 71, p. 505-509, 1991.

EMPRESA BRASILEIRA DE PESQUISA AGROPECUARIA. Available at: <http://sistemasdeproduçào.cnptia.embrapa.br/FONTESHTML/Mandioca/Mand ioca centrosul/importância.htm>. Accessed on: January 15, 2010.

FALLER, A. L. K.; FIALHO, E. Availability of polyphenols in fruits and vegetables consumed in

Brazil. **Revista de Saùde Pùblica**, Sao Paulo, v. 43, n. 2, p. 211-218, 2009.

FERREIRA, D. F. **SISVAR**: verao 4.6 (build 61) software. Lavras: uFLA, 2003. Available at: <htpp://www.dex.ufla.br/danielff/dff02.htm>. Accessed on: January 15, 2010.

GIDAMIS, A. B.; BRIEN, G. M.; POuLTER, N. H. Cassava detoxification of traditionalTanzanian cassava foods. **International Journal of Food Science and Technology,** Oxford, v. 28, n. 2, p. 211-218, Apr. 1993.

HALLIWELL, B. Antioxidants in human health and disease. **Annual Review of Nutrition**, Palo Alto, v. 16, p. 33-50, 1996.

HALLIWELL, B. Free radicals and antioxidants: a personal view. **Nutrition Reviews**, New York, v. 52, n. 8, p. 253-265, 1994.

HALLIWELL, B. Free radicals and other reactive species in disease. In: ENCYCLOPEDIA of life sciences. Coimbra: Nature Publishing Group, 2001. p. 1-7.

HALLIWELL, B.; GuTTERIDGE, J. M. C.; CROSS, C. E. Free radicals, antioxidants, and human disease: where are we now ? **Journal of Laboratory and Clinical Medicine**, Saint Louis, v. 119, p. 598, 1992.

HALLIWELL, B.; GUTTERIDGE, J. M. C. **Free radicals in biology and medicine**. 3. ed. New York: Oxford Science Publications, 1999. 936 p.

HENDRY, G. A. Plant pigments. In: LEA, P. J.; LEEGOOD, R. C. (Ed.). **Plant biochemistry and molecular biology**. Great Britain: Bookcraft, 1993. p. 181196.

HERNANDEZ, Y.; LOBO, M. G.; GONZALEZ, M. Determination of vitamin C in tropical fruits: a comparative evaluation of methods. **Food Chemistry**, London, v. 96, n. 4, p. 654-664, Aug. 2006.

HUANG, D.; OU, B.; PRIOR, R. L. The chemistry behind antioxidant capacity assays. **Journal of Agricultural and Food Chemistry**, London, v. 53, p. 19811856, 2005.

INNIS, S. H. Perinatal biochemistry and physiology oflong-chain polyunsaturated fatty acids. **Jornal de Pediatria,** Rio de Janeiro, v. 143, p. 1-8, 2003.

BRAZILIAN Institute of Geography AND STATISTICS. Available at: <http://www.sectec.go.gov.br/links/institutos/ ibge.htm>. Accessed on: 15 Jan. 2010.

JOHANNING, G. L. Modulation of breast cancer cell adhesion by unsaturated fatty acids. **Nutrition**, Alabama, v. 12, p. 810-866, 1996.

JOSEPH, J. D.; ACKMAN, R. G. Capillary column gas chromatography method for analysis of encapsulated fish oil and fish oil ethyl esters: collaborative study.

Journal of AOAC International, Arlington, v. 75, p. 488-506, 1992.

KAUR, C.; KAPOOR, H. C. Anti-oxidant activity and total phenolic content of some Asian vegetables. **International Journal of Food Science and Technology**, Oxford, v. 37, p. 153-161, 2002.

KUBOLA, J.; SIRIAMORNPUM, S. Phenolic contents and antioxidant activities of bitter gourd (Momordica Charantia L.) leafs stem and fruit fraction extracts in vitro. **Food Chemistry**, London, v. 1110, n. 4, p. 881-890, Oct. 2008.

KUSKOSKI, E. M.; ASUERO, A. G.; TRONCOSO, A. N.; FILHO, J. M.;

FETT, R. Aplicación de diversos métodos quimicos para determinar a actividad antioxidant en pulpa de frutos. **Ciência e Tecnologia dos Alimentos**, Campinas, v. 25, n. 4, p. 726-732, Oct./Dec. 2005.

LEHNINGER, A. L.; NELSON, D. L.; COX, μ. M. **Lehninger principles of biochemistry**. 4. ed. Sao Paulo: Sarvier, 2006. 1202 p.

LEONG, L. P.; SHUI, G. An Investigation of antioxidant capacity fruits in Singapure markets. **Food Chemistry**, Washington, v. 76, p. 69-75, 2002.

LIMA, A. J. B. **Characterization and antioxidant activity of jaboticaba**. 2009. 159 p. Thesis (Doctorate in Agrochemistry) - Federal University of Lavras, Lavras.

LIMA, E. S.; ABDALLA, D. S. P. Lipid peroxidation: mechanisms and evaluation in biological samples. **Revista Brasileira de Ciências Farmacêuticas**, Sao Paulo, v. 37, n. 3, Sep./Dec. 2001.

LORENZI, J. A. **Cassava.** Campinas: CATI, 2003 (Technical bulletin).

LOREZI, J. O.; DIAS, C. A. C. **Cassava culture**. Campina: CATI, 1993. 41 p. (CATI technical bulletin, n. 211).

MACHLIN, L. J.; BENDICH, A. Free radical tissue damage: protective role of antioxidant nutrients. **The FASEB Journal**, Bethesda, v. 1, n. 2, p. 441-445, 1987.

MARCO, G. A rapid method for evaluation of antioxidants. **Journal of the American Oil Chemists' Society**, Illinois, v. 45, p. 594-598, 1968.

MAILLARD, V.; BOUGNOUX, P.; FERRARI, P.; JOURDAN, M.; PINAULT, M.; LAVILLONNIÈRE, F. N-3 and n-6 fatty acids in breast cancer adipose tissue and relative risk of breast cancer in a case-control study in Tours, France. **International Journal of Cancer**, New York, v. 98, p. 78-83, 2002.

MELO, D. S. de; CORRÊA, A. D.; MARCOS, F. C. A.; SOUSA, R. V. de; ABREU, C. M. P. de;

SANTOS, C. D. dos. Effects of cassava leaf flour on lipid peroxidation, blood lipid profile and liver weight in rats. **Ciência e Agrotecnologia**, Lavras, v. 31, n. 2, p. 420-428, mar./abr. 2007.

MELO, E. A.; MACIEL, M. I. S.; LIMA, V. L. A. G.; ARAUJO, C. R. Total phenol content and antioxidant capacity of frozen fruit pulps. **Alimentos e Nutriçâo**, Marilia, v. 19, n. 1, p. 67-72, jan./mar. 2008a.

MELO, E. A.; MACIEL, M. I. S.; LIMA, V. L. A. G.; NASCIMENTO, R. J. Antioxidant capacity of fruits. **Revista Brasileira de Ciências Farmacêuticas**, Sao Paulo, v. 44, n. 2, Apr./Jun. 2008b.

MINOLTA CAMERA. **Manual for chlorophyll meter SPAD 502**. Osaka: Minolta, Radiometric Instruments Divisions, 1989. 22 p.

MORAES, F. P.; COLLA, L. M. Functional foods and nutraceuticals: definitions, legislation and health benefits. **Revista Eletrônica de Farmácia**, Goiânia, v. 2, p. 109-122, 2006.

NAGATA, M.; YAMASHITA, I. Simple method for simultaneous determination of chlorophyll and carotenoids in tomato fruit. **Nippon Shokuhin Kogyo Gakkaishi**, Tokyo, v. 39, n. 10, p. 925-928, 1992.

NEVES, O. S. C.; CARVALHO, J. C.; MARTINS, F. A. D.; PADUA, T. R. P.; PINHO, P. J. Use of SPAD-502 in the evaluation of chlorophyll, nitrogen, sulphur, iron and manganese levels in herbaceous cotton. **Pesquisa Agropecuâria Brasileira**, Brasilia, v. 40, n. 5, p. 75-78, May 2005.

ORTEGA-FLORES, C. I.; COSTA, M. I.; CEREDA, M. P.; PENTEADO, M. V. C. Evaluation of the protein quality of dehydrated cassava leaves (*Manihot esculenta* Crantz). **Nutrire**, Sao Paulo, v. 25, n. 1, p. 47-59, jun. 2003.

OTSUBO, A. A. **Cultivation of cassava in the Center-South region of Brazil**. Dourados: EMBRAPA Agropecuâria Oeste, 2004.

PADJMA, G. Evaluation of techniques to reduce assayable tannim and cyanide in cassava leaves. **Journal of Agricultural and Food Chemistry**, London, v. 37, p. 712-716, 1989.

PEREIRA, G. I. S.; PEREIRA, R. G. F. A.; BARCELOS, M. F. P.; MORAIS, A. R. Chemical evaluation of carrot leaves for use in human food. **Ciência e Agrotecnologia**, Lavras, v. 27, n. 4, p. 852-857, jul./ago. 2003.

RAMALHO, V. C.; JORGE, N. Antioxidants used in oils, fats and fatty foods. **Quimica Nova**, Sao Paulo, v. 29, n. 4, p. 755-760, 2006.

RE, R.; PELLEGRINI, N.; PROTEGGENTE, A.; PANNALA, A.; YANG, M.; RICE-EVANS, C. Antioxidant activity applying an improved ABTS radical cation decolorization assay. **Free Radical Biology and Medicine**, New York, v. 26, n. 9/10, p. 1231-1237, 1999.

RIBEIRO, E. P.; SERAVALLI, E. A. G. **Quimica de alimentos**. Sao Paulo: E. Blücher, 2004. 157 p.

RODRIGUES, H. G.; DINIZ, Y. S.; FAINE, L. A.; ALMEIDA, J. A.; FERNANDES, A. A. H.; NOVELLI, E. L. B. Nutritional supplementation with natural antioxidants: effect of rutin on HDL-cholesterol concentration, **Revista de Nutriçâo,** Campinas, v. 16, n. 3, jul./set. 2003.

ROVER JÛNIOR, L.; HOEHR, N. F.; VELLASCO, A. P. Antioxidant system involving the glutathione metabolic cycle associated with electroanalytical methods in the evaluation of stress. **Quimica Nova**, Sao Paulo, v. 24, n. 1, p. 112-119, 2001.

RUFINO, M. S. M.; ALVES, R. S.; BRITO, E. S.; MORAIS, S. M. **Determination of total antioxidant activity in fruit by ABTS free radical scavenging^{+}** . Fortaleza: EMBRAPA, 2007.

SANCHEZ-MORENO, C. Compuestos polifenólicos: efectos fisiológicos: actividad antioxidante. **Alimentaria**, Lisbon, v. 1, p. 29-40, 2002.

SANTOS, L. C.; PIACENTE, S.; MONTORO, P.; PIZZA. C. Antioxidant activity of xanthones isolated from Leiothrix species. **Revista Brasileira de Farmacognosia**, Curitiba, v. 13, n. 2, p. 64-67, 2003.

SARTORELLI, C. S. do C. **Chemical characterization of the aerial part of carrots and beets, with a view to their use in human food**. 1998. 98 p. Dissertation (Master's Degree in Food Sciences) - Federal University of Lavras, Lavras.

SIES, H. Strategies of antioxidant defense. Review. **European Journal of Biochemistry**, Berlin, v. 215, n. 2, p. 213- 219, 1993.

SILVA, P. A.; QUEIROZ, E. R; ABREU, C. M. P.; SACZK, A. A. Development of analytical methodology for the determination of vitamin C in strawberries by HPLC. In: REUNIÂO ANUAL DA SOCIEDADE BRASILEIRA DE QUiMICA, 32., 2009, Fortaleza. **Anais...** Fortaleza: [s.n.], 2009. 1 CDROM.

SILVA, S. R.; MERCADANTE, A. Z. Carotenoid composition of yellow passion fruit *(Passiflora edulis flavicarpa) In Natura*. **Ciência e Tecnologia de Alimentos**, Campinas, v. 2, p. 254-258, Sep./Dec. 2002.

SIMOPOULOS, A. P. Omega-3 fatty acids in wild plants, nuts and seeds. **Journal of Clinical Nutrition**, Philadelphia, v. 11, n. 6, p. 163-73, 2002.

SOARES, M.; WELTER, L.; KUSKOSKI, E. M.; GONZAGA, L.; FETT, R. Phenolic compounds and antioxidant activity of Niagara and Isabel grape skins.

Revista Brasileira de Fruticultura, Cruz das Almas, v. 30, n. 1, p. 59-64, mar. 2008.

SOARES, S. E. Phenolic acids as antioxidants. **Revista de Nutriçâo,** Campinas, v. 15, n. 1, p. 71-81, 2002.

STATISTICAL ANALYSIS SYSTEM. **Users quide**: statistcs : version 6. 4. ed.

Cary, 1999. v.2, 1686 p.

STROHECKER, R.; HENNING, H. M. **Analisis de vitaminas**: metodos comprobados. Madrid: Paz Montalvo, 1967. 428 p.

TAVARES, J. T. Q.; SILVA, C. L.; CARVALHO, L. A.; SILVA, M. A.;

SANTOS, C. M. G. Stability of ascorbic acid in orange juice submitted to different treatments. **Magistra**, Cruz das Almas, v. 12, n. 1/2, 2000. Not paginated.

TURATTI, J. M.; GOMES, R. A. R.; ATHIÉ, I. **Lipideos***:* functional aspects and new trends. Campinas: ITAL, 2002. 78 p.

UENOJO, M.; JUNIOR, M. R. M.; PASTORE, G. M. Carotenoids: Properties, applications and biotransformation to form aroma compounds. **Quimica Nova**, Sao Paulo, v. 30, n. 3, p. 616-622, 2007.

WASKOM, R. M.; WESTFALL, D. G.; SPELLMAN, D. E.; SOLTANPOUR, P. N. Monitoring nitrogen status of corn with a portable chlorophyll meter.

Communications in Soil Science and Plant Analysis, New York, v. 27, p. 545560, 1996.

WOBETO, C.; CORRÊA, A. D.; ABREU, C. M. P. de; SANTOS, C. D;

ABREU, J. R. Nutrients in the cassava (*Manihot esculenta* Crantz) leaf powder at three ages of the plant. **Ciência e Tecnologia de Alimentos**, Campinas, v. 26, n. 4, p. 865-869, Oct./Dec. 2006.

WOBETO, C.; CORRÊA, A. D.; ABREU, C. M. P. de; SANTOS, C. D.;

PEREIRA, H. V. Antinutrients in the cassava (*Manihot esculenta* Crantz) leaf powder at three ages of the plant. **Ciência e Tecnologia de Alimentos**, Campinas, v. 27, n. 1, p. 108-112, jan./mar. 2007.

WYNDER, E. L.; COHEN, L.; MUSCAT, J. E.; WINTERS, B.; DWYER, J. T.; BLACKBURN, G. Breast cancer: weighing the evidence for a promoting role of dietary fat. **Journal of the National Cancer Institute**, Bethesda, v. 89, p. 76675, 1997.

YILDIRIM, A.; MAVI, A.; KARA, A. A. Determination of antioxidant and antimicrobial activities of Rumex crispus L. extracts. **Journal of Agricultural and Food Chemistry**, Chicago, v. 49, p.

4083-4089, 2001.

ZHENG, W.; WANG, S. Y. Antioxidant activity and phenolic compounds in selected herbs. **Journal of Agricultural and Food Chemistry**, Chicago, v. 49, p. 5165-5170, 2001.

ZOTARELLI, L.; CARDOSO, E. G.; PICCINI, J. L.; URQUIAGA, S.;

BODDEY, R. M.; TORRES, E.; ALVES, B. J. R. Calibration of the Minolta SPAD-502 chlorophyll meter for evaluating the nitrogen content of maize. **Pesquisa Agropecuâria Brasileira**, Brasilia, v. 38, p. 1-5, 2003.

7. ANNEX

TABLE 1 Summary of the analysis of variance for leaf moisture and FFM of four cultivars and chlorophyll in TIP.

		QM		
FV	GL	Leaf humidity	Humidity FFM	Chlorophyll
C	3	24,16**	2,88 ^{ns}	244363,26**
I	2	97,28**	32,21**	39114,07**
C x I	6	5,42 ^{ns}	0,69 ^{ns}	5313,04**
Waste	24	2,63	1,33	67,76
CV (%)		2,28	15,82	1,35

NS/** Non-significant F-test, significant at 1% probability, respectively.

TABLE 2 Summary of the analysis of variance for the solvent tests.

FV	GL	QM
Solvent	8	52712,87**
Waste	18	57,72
CV (%)	2,96	

** F test significant at 1% probability.

TABLE 3 Summary of the analysis of variance of the antioxidant substances and antioxidant tests of the FFM of four cultivars, in TIP.

		QM				
FV	GL	Vitamin-C	Polyphenols	β-carotene	ABTS ά	β-carotene/ linoleic acid
Cultivar (C)	3	11375,06**	491,94**	73,16**	89798,76**	197,98**
Age (I)	2	258135,56**	1484,25**	217,77**	340843,39*	* 65,15**
C x I	6	2593,88**	21,75**	282,28**	15355,57**	4,47**
Waste	24	208,04	0,71	6,85	1,24	1,73
CV (%)		4,90	2,30	4,68	0,29	1,54

** F test significant at 1% probability.

MIX
Papier aus verantwortungsvollen Quellen
Paper from responsible sources
FSC® C105338

Printed by Books on Demand GmbH, Norderstedt / Germany